DATE DUE

JY 21 '97			

DEMCO 38-296

Fatty Acid and Lipid Chemistry

Fatty Acid
and
Lipid Chemistry

F.D. GUNSTONE
Honorary Research Professor
Lipid Chemistry Unit
School of Chemistry
University of St Andrews
Fife, UK

BLACKIE ACADEMIC & PROFESSIONAL
An Imprint of Chapman & Hall

London · Glasgow · Weinheim · New York · Tokyo · Melbourne · Madras

print of Chapman & Hall,
lasgow G64 2NZ

Chapman & Hall, 2–6 Boundary Row, London SE1 8HN, UK

Blackie Academic & Professional, Wester Cleddens Road, Bishopbriggs, Glasgow G64 2NZ, UK

Chapman & Hall GmbH, Pappelallee 3, 69469 Weinheim, Germany

Chapman & Hall USA, Fourth Floor, 115 Fifth Avenue, New York NY 10003, USA

Chapman & Hall Japan, ITP-Japan, Kyowa Building, 3F, 2-2-1 Hirakawacho, Chiyoda-ku, Tokyo 102, Japan

DA Book (Aust.) Pty Ltd, 648 Whitehorse Road, Mitcham 3132, Victoria, Australia

Chapman & Hall India, R. Seshadri, 32 Second Main Road, CIT East, Madras 600 035, India

First edition 1996

© 1996 Chapman & Hall

Typeset in 10/12pt Times by Cambrian Typesetters, Frimley, Surrey

Printed in Great Britain by Hartnoll's Ltd, Bodmin

ISBN 0 7514 0253 2

∞ Printed on acid-free text paper, manufactured in accordance with ANSI/ NISO Z39.48-1992 (Permanence of Paper)

Preface

This book has a pedigree. It has developed from earlier publications by the author and from his experience over 50 years in reading, writing, thinking, and working with lipids and fatty acids. The earlier publications are:

(i) *An Introduction to the Chemistry of Fats and Fatty Acids*, Chapman and Hall, 1958.

(ii) *An Introduction to the Chemistry and Biochemistry of Fatty Acids and their Glycerides*, Chapman and Hall, 1967.

(iii) *Lipids in Foods: Chemistry, Biochemistry, and Technology* (with F.A. Norris), Pergamon Press, 1983.

(iv) *The Lipid Handbook* (with J.L. Harwood and F.B. Padley), Chapman and Hall, first edition 1986, second edition 1994.

(v) *A Lipid Glossary* (with B.G. Herslof), The Oily Press, Dundee, 1992.

(vi) Lecture notes for a course on *Fatty Acids and Lipids* designed for those entering the oil and fat industry and given on over 20 occasions since 1977.

The book is dedicated to the next generation of lipid scientists. The study of lipids now involves many disciplines, all of which require a basic knowledge of the chemical nature and properties of these molecules, which is what this book is about. It is written particularly for those who, with some knowledge of chemistry or biochemistry, need to know more about the nature of lipids and fatty acids. I expect that many of my readers will be employed in the food industry since 80% of the world production of oils and fats is eaten by humans and another 6% goes into animal feed. They will need to understand the materials they handle; their origin, their chemical nature, the effects of processing, and their physical, chemical, biochemical, and nutritional properties. Another group of readers will be employed in the oleochemical industry, modifying the material produced by nature for the benefit of human kind. They will have to understand the constraints of production and of chemistry within which they work and they will have to be aware of the present state of knowledge of these materials. Another group may consider themselves to be academic researchers but they cannot escape from the real world of marketplace availability, and will need to know something about sourcing, about the changes which occur on refining oils and fats and how these materials can be modified on a commercial scale.

For the most part, I have given only general references at the end of each chapter. These will furnish detailed information for those who need it. Not surprisingly, my first port of call would be *The Lipid Handbook* which, in its second edition, has 722 pages of text and an extensive selection of dictionary entries with references, abstracted from the *Dictionary of Organic Compounds* (551 pages).

In preparing a book of this limited size, I have had to be selective and although some less common fatty acids and some of the less useful chemistry of the acids has not been completely ignored, I have nevertheless emphasized throughout the most common and the useful features of fatty acid and lipid chemistry. My appreciation of these matters is based on long experience, wide reading, attendance at many conferences and frequent contact with those who work in the food and oleochemical industry. Apart from a few historical comments I have dwelt on things as they now are. There are, of course, new developments but it is too soon to fully appreciate their significance. Some research reports will gather dust on library shelves and become archive material; others will be the seed from which there will be significant growth. I have had to use my own judgement in deciding between these; if I have got it right more often than I have got it wrong I have done well.

The book is divided into ten chapters. It is necessary first to describe the fatty acids (chapter 1) and the lipids (chapter 2) and to discuss their nomenclature, structure, biosynthesis, and chemical synthesis. This is followed by an account of the major sources of oils, fats, and other lipids (chapter 3), a description of the processes by which such materials are made available and are modified (chapter 4), and then the analytical procedures used for their investigation (chapter 5). The physical and chemical properties of the fatty acid and lipids are covered in the next three chapters and the final two chapters are devoted to nutrition and to the major food and non-food uses of lipids.

The book has been read in an earlier draft form by Dr F.B. Padley (Unilever) and by Dr H. Eierdanz (Henkel) and I acknowledge the assistance they have given me in weeding out errors and pointing out omissions. The book is better for their contributions but they hold no responsibility for any remaining errors.

F.D.G.

Contents

1 **Fatty acids – Nomenclature, structure, isolation and structure-determination, biosynthesis, and chemical synthesis**　　**1**

  1.1　Fatty acid nomenclature　　1
  1.2　Fatty acids – main structural features　　3
  1.3　Saturated acids　　4
    1.3.1　Short- and medium-chain saturated acids (C_4–C_{14})　　4
    1.3.2　Palmitic and stearic acids　　5
    1.3.3　Long-chain acids　　6
  1.4　Monoene acids　　6
  1.5　Methylene-interrupted polyene acids　　8
  1.6　Conjugated polyene acids　　9
  1.7　Other polyene acids　　10
  1.8　Acetylenic and allenic acids　　10
  1.9　Acids with *trans*-olefinic unsaturation　　11
  1.10　Branched-chain acids　　12
  1.11　Cyclic acids　　12
  1.12　Oxygenated acids　　13
    1.12.1　Hydroxy acids　　13
    1.12.2　Epoxy acids　　15
    1.12.3　Furanoid acids　　15
  1.13　Other acids　　16
  1.14　Fatty acids – isolation and identification　　16
    1.14.1　Thin-layer chromatography　　18
    1.14.2　High-performance liquid chromatography　　20
    1.14.3　Gas chromatography　　20
    1.14.4　Spectroscopic procedures of identification　　22
    1.14.5　Chemical procedures of identification　　22
  1.15　Fatty acid biosynthesis　　23
    1.15.1　*de novo* Synthesis of saturated acids　　23
    1.15.2　Chain elongation　　26
    1.15.3　Desaturation to monoene acids　　27
    1.15.4　Desaturation to polyene acids　　27
  1.16　Fatty acids – chemical synthesis　　28
    1.16.1　Introduction　　28
    1.16.2　Synthesis *via* acetylenic intermediates　　30
    1.16.3　Synthesis by the Wittig reaction　　31
    1.16.4　Isotopically labelled acids　　32
  Bibliography　　33

2 **Lipids – Nomenclature, structure, biosynthesis, and chemical synthesis**　　**35**

  2.1　Introduction　　35
  2.2　Acylglycerols　　36
    2.2.1　Monoacylglycerols　　37
    2.2.2　Diacylglycerols　　37
    2.2.3　Triacylglycerols　　37

2.3	Glycosyldiacylglycerols	39
2.4	Phosphoglycerides	39
2.5	Ether lipids	42
2.6	Acyl derivatives of other alcohols	43
2.7	Sphingolipids	43
2.8	Lipid biosynthesis	44
2.9	Chemical synthesis of acylglycerols and phospholipids	45
	2.9.1 Introduction	45
	2.9.2 Acylation	47
	2.9.3 Protecting groups	48
	2.9.4 C_3 compounds other than glycerol	48
	2.9.5 Synthetic procedures for racemic acylglycerols	49
	2.9.6 Synthetic procedures for enantiomeric acylglycerols	52
	2.9.7 Structured lipids	54
	2.9.8 Synthesis of phospholipids	57
	Bibliography	59

3 The major sources of oils, fats, and other lipids — **61**

3.1	Fats and oils in the marketplace	61
3.2	The major vegetable oils and fats	64
	3.2.1 Blackcurrant, borage, and evening primrose oils	64
	3.2.2 Castor oil	65
	3.2.3 Cocoa butter	65
	3.2.4 Coconut oil	65
	3.2.5 Corn oil	66
	3.2.6 Cottonseed oil	66
	3.2.7 Gourmet oils	66
	3.2.8 Groundnut oil	67
	3.2.9 Linseed oil	68
	3.2.10 Olive oil	68
	3.2.11 Palm oil and palmkernel oil	68
	3.2.12 Rapeseed oil	68
	3.2.13 Rice bran oil	69
	3.2.14 Safflower oil	70
	3.2.15 Sesame oil	70
	3.2.16 Soybean oil	70
	3.2.17 Sunflower oil	70
	3.2.18 Tall oil	71
3.3	Fats of animal origin	71
	3.3.1 Butter fat	71
	3.3.2 Lard and animal tallows	71
	3.3.3 Fish oils	72
3.4	Fatty acid composition	73
3.5	Triacylglycerol composition	73
3.6	Interchangeability of oils and fats	76
3.7	New oilseed crops	77
	3.7.1 Lauric oils	78
	3.7.2 Oleic oils	78
	3.7.3 Oils containing petroselinic acid	79
	3.7.4 Oils containing higher monoene acid (C_{20}–C_{24})	79
	3.7.5 Oils containing C_{18} polyene acids	80
	3.7.6 Oils containing epoxy acids	81
	3.7.7 Oils containing hydroxy acids	81
3.8	Phospholipids: sources and composition	81
3.9	Minor components	82
3.10	Waxes	84
	Bibliography	85

4 Processing: Extraction, refining, fractionation, hydrogenation, interesterification   **87**

4.1 Extraction   87
4.2 Refining   88
    4.2.1 Degumming   90
    4.2.2 Neutralization   90
    4.2.3 Bleaching   90
    4.2.4 Deodorization or physical refining   90
    4.2.5 Chromatographic refining (super refining)   90
4.3 Hydrogenation   91
4.4 Fractionation   95
4.5 Interesterification   96
Bibliography   98

5 Analytical procedures   **100**

5.1 Standard procedures   100
    5.1.1 Sampling and oil content   100
    5.1.2 Melting behaviour   101
    5.1.3 Unsaturation   102
    5.1.4 Acidity, saponification, unsaponifiable, hydroxyl value   102
    5.1.5 Oxidative deterioration and oxidative stability   103
5.2 Lipid analysis   104
    5.2.1 The nature of the problem   104
    5.2.2 Lipid classes   106
    5.2.3 Component acids (alcohols etc.)   106
    5.2.4 Triacylglycerols   111
    5.2.5 Phospholipids   117
    5.2.6 Stereospecific processes   118
    5.2.7 Combined processes   125
Bibliography   128

6 Physical properties   **129**

6.1 Polymorphism, crystal structure, and melting point   129
    6.1.1 Introduction   129
    6.1.2 Alkanoic acids   130
    6.1.3 Glycerol esters   133
    6.1.4 Margarines and confectionery fats   137
6.2 Ultraviolet spectroscopy   137
6.3 Infrared spectroscopy   138
6.4 Electron spin resonance (ESR) spectroscopy   139
6.5 ^{1}H NMR spectroscopy   140
    6.5.1 Low-resolution spectroscopy   140
    6.5.2 High-resolution spectroscopy   142
6.6 ^{13}C NMR spectroscopy   143
6.7 Mass spectrometry   147
    6.7.1 Derivatives of olefinic compounds   149
    6.7.2 Nitrogen-containing esters and amides   151
    6.7.3 Tandem mass spectrometry   153
Bibliography   155

7   Reactions associated with double bonds   157

   7.1   Catalytic hydrogenation, chemical reduction, biohydrogenation   157
      7.1.1   Catalytic hydrogenation   157
      7.1.2   Other chemical reductions   159
      7.1.3   Biohydrogenation   161
   7.2   Autoxidation and photo-oxygenation   162
      7.2.1   Introduction   162
      7.2.2   Autoxidation   162
      7.2.3   Photo-oxygenation   164
      7.2.4   Decomposition of hydroperoxides to short-chain compounds   165
      7.2.5   Reaction products   167
      7.2.6   Other secondary reaction products   171
   7.3   Antioxidants   173
      7.3.1   Introduction   173
      7.3.2   Primary antioxidants (chain-breaking)   175
      7.3.3   Secondary antioxidants   179
      7.3.4   Other materials which inhibit oxidative deterioration   179
   7.4   Biological oxidation   180
      7.4.1   α-Oxidation   180
      7.4.2   β-Oxidation   181
      7.4.3   ω-Oxidation   181
      7.4.4   Lipoxygenase   182
      7.4.5   Production and function of eicosanoids   184
   7.5   Other oxidation reactions   186
      7.5.1   Epoxidation   186
      7.5.2   Hydroxylation   189
      7.5.3   Oxidative fission   189
   7.6   Halogenation   192
   7.7   Oxymercuration   195
   7.8   Stereomutation   196
   7.9   Metathesis   197
   7.10 Double bond migration and cyclization   198
   7.11 Dimerization   200
   7.12 Other double bond reactions   201
   Bibliography   204

8   Reactions of the carboxyl group   205

   8.1   Introduction   205
   8.2   Hydrolysis   206
   8.3   Esterification   207
      8.3.1   Esterification between acids (and related acyl derivatives) and
            alcohols   208
      8.3.2   Acidolysis: reaction between esters and acids   209
      8.3.3   Alcoholysis: reaction between esters and alcohols (methanolysis,
            glycerolysis)   209
      8.3.4   Interesterification: reaction between esters and esters   210
      8.3.5   Structured lipids   213
   8.4   Acid chlorides, anhydrides, and ketene dimers   214
   8.5   Peroxy acids and esters   215
   8.6   Alcohols   217
   8.7   Nitrogen-containing compounds   218
      8.7.1   Amides   218
      8.7.2   Nitriles, amines, and their derivatives   218
   8.8   Other reactions of the carboxyl group   220
   Bibliography   222

9   Dietary fats and nutrition 223

9.1	Storage and structural fats	223
9.2	Digestion, absorption, and transport	224
9.3	The role of fats in health and disease	225
	Bibliography	228

10   Practical applications of oils and fats 229

10.1	Introduction	229
10.2	Butter	229
10.3	Margarine	232
10.4	Other food uses	234
10.5	Food-grade surfactants and emulsifiers	234
10.6	Non-food uses of fatty acids and their derivatives	236
	Bibliography	242

Index 244

1 Fatty acids – Nomenclature, structure, isolation and structure determination, biosynthesis and chemical synthesis

There is no generally accepted definition of the class of natural products designated as lipids and the author continues to use his earlier definition that "lipids are compounds based on fatty acids or closely related compounds such as the corresponding alcohols and the sphingosine bases". On this basis it is reasonable to start this book with a discussion of the fatty or long-chain acids which are an essential part of most lipids. The chemical structures of these acids and their physical, chemical and biological properties are basic to the understanding of the physics, chemistry, and biochemistry of lipids. In a few lipids the acids are replaced by related long-chain compounds such as alcohols, aldehydes, or amines.

1.1 Fatty acid nomenclature

Fatty acids are designated in several different ways. Despite the alternative descriptions set out here many trivial names are still widely used and have therefore to be known. These names were originally given before the chemical structure of the acid was elucidated and were often chosen to indicate the source of the acid. Examples include: palmitic (from palm oil), oleic (from olive oil, *Olea europea*), linoleic and linolenic (from linseed oil), ricinoleic (from castor oil, *Ricinus communis*), and arachidic acid (from groundnut oil, *Arachis hypogea*). Less often the name is linked with the scientist who first described the acid or some significant property of that acid, e.g. Mead's acid (20:3 *n*-9). Sometimes an acid had more than one trivial name until the identity of the material from different sources was established. Thus 9-hexadecenoic acid was earlier designated palmitoleic acid and also zoomaric acid. A trivial name may also be assigned or may continue to be used because the systematic name is cumbersome as for example with α-eleostearic acid which is simpler than 9*c*11*t*13*t*-octadecatrienoic acid. Trivial names are easy to use but they are not, in themselves, indicative of structure; that has to be remembered as well as the name.

Systematic names are based on internationally accepted rules agreed by organic chemists and biochemists. Those who know the rules can interconvert systematic names and structures. As a simple example oleic acid is *cis*-9-octadecenoic acid. This is a carboxylic acid (oic) with 18 carbon atoms (octadec) and one olefinic centre (en) which lies between carbon 9 and 10 (counting from the carboxyl end) and has *cis* configuration, i.e.

$$CH_3(CH_2)_7CH=CH(CH_2)_7COOH$$

which may also be represented by the line drawing:

Representations like this are increasingly popular. They are more useful when the number of double bonds and/or other functional groups is larger and the saturated sections of the molecule are short. The structure shown below is not easily recognizable as stearic acid (octadecanoic) as the number of carbon atoms represented in this structure is not immediately apparent and has to be carefully counted:

This difficulty is overcome in the alternative formulation:

Even though the line forms of these structures are useful because of their immediate visual impact, they are not convenient for tabulated data or for insertion into lines of text. Because the word form can be complex and clumsy, systematic or trivial names are sometimes abbreviated to two or three capital letters as in the following examples:

gamma-linolenic acid   GLA
arachidonic acid   AA
eicosapentaenoic acid   EPA
docosahexaenoic acid   DHA

The third way of designating fatty acids involves the use of numbers such as 18:2. This symbol describes an acid such as linoleic with 18 carbon atoms (assumed to be straight-chain) and two unsaturated centres (assumed to be *cis*-olefinic). Since there are many isomeric compounds which could be represented by this symbol additional descriptors may be added thus:

18:2 (9,12), 18:2 (9*c*,12*c*), 18:2 (9Z,12Z), 18:2 (*n*-6)

All these refer to the same acid. The first indicates the position of the two unsaturated centres in the C_{18} chain with reference to COOH = 1. The second and third formulations confirm the *cis* or *Z* configuration of the double bonds. The symbol Δ is sometimes added to show that numbering is with respect to the acid function. The fourth designation introduces a further concept in fatty acid nomenclature.

Sometimes it is useful to designate double bond positions with respect to the end CH_3 group and this is done with symbols such as ω6 or *n*-6 which indicate that the first double bond is on carbon 6 counting from the methyl group. In the absence of other information it is assumed that double bonds are methylene-interrupted and have *cis* (*Z*) configuration. Symbols such as *c*, *t*, *e* are used to show *cis*, *trans*, and ethylenic unsaturation (where configuration is not known or does not apply) and *a* (acetylenic) or *y* (ynoic) is used to show a triple bond.

All of these systems of nomenclature will be used as appropriate throughout this text and the lipid scientist must expect to meet them all.

1.2 Fatty acids – main structural features

The number of known natural fatty acids exceeds 1000 although only a relatively small number – perhaps 20–50 – are of common concern. Based on a survey of these 1000 structures, and noting particularly the structures of those acids produced most commonly in nature, it is possible to make four general statements. Each of these is generally true but there are exceptions to all four. The exceptions are frequently trivial but sometimes they are significant. Though originally based on a survey of chemical structures it is now clear that these statements also reflect the underlying biosynthetic pathways by which the acids are produced in nature (section 1.15).

(i) Natural fatty acids – both saturated and unsaturated – are straight-chain compounds with an even number of carbon atoms in their molecules. This is true for the great majority of structures and for the more abundant acids. Chain lengths range from two to over 80 carbon atoms although they are most commonly between C_{12} and C_{22}. Despite the validity of this statement, acids with an odd number of carbon atoms (e.g. heptadecanoic, C_{17}) occur as do those with branched chains (e.g. isopalmitic, ante-isononadecanoic) or with carbocyclic units (e.g. sterculic, chaulmoogric).

(ii) Acids with one unsaturated centre are usually olefinic compounds with *cis* (*Z*) configuration and with the double bond in one of a limited number of preferred positions. This is most commonly Δ9 (i.e. nine carbon atoms from the carboxyl group as in oleic) or *n*-9 (i.e. nine carbon atoms from the methyl group as in oleic or erucic acid). But double bonds occur in other positions (e.g. petroselinic, 6c-18:1), or have *trans* configuration (e.g. elaidic 9t-18:1), or can be replaced by an acetylenic unit (e.g. tariric 6a-18:1).

(iii) Polyunsaturated acids are mainly polyolefinic with a methylene-interrupted arrangement of double bonds having *cis* (*Z*) configuration. That is, *cis* double bonds are separated from each other by one CH_2 group as in arachidonic acid:

The 1,4-pattern of unsaturation is characteristic of natural fatty acids and differs from that in acyclic isoprenoids which is usually 1,3 (conjugated) or 1,5. Polyunsaturated acids occur in biosynthetically related families. The most important are the *n*-6 acids based on linoleic and the *n*-3 acids based on α-linolenic acid. In contrast to this very common pattern of unsaturation some acids have conjugated unsaturation which may be *cis* or *trans* (e.g. eleostearic, calendic, parinaric acids), some have mixed en/yne unsaturation which may be conjugated (e.g. isanic) or non-conjugated (e.g. crepenynic), and some have non-conjugated unsaturation which is not entirely methylene-interrupted. These are known as non-methylene-interrupted polyenes (e.g. columbinic, pinolenic).

(iv) Fatty acids rarely have functional groups apart from the carboxyl group and the various types of unsaturation already discussed. Nevertheless acids are known which also contain a fluoro, hydroxy, keto, or epoxy group. Two important examples are ricinoleic (12-hydroxyoleic acid) and vernolic acid (12,13-epoxyoleic).

These generalizations have a biosynthetic basis and even some of the exceptions are accommodated into the general biosynthetic scheme with only minor modifications.

Based on the annual production of commercial vegetable oilseeds it has been estimated that eight acids account for about 97% of the total production: lauric (4%), myristic (2%), palmitic (11%), stearic (4%), oleic (34%), linoleic (34%), α-linolenic (5%), and erucic (3%). The level of linolenic would rise if all green tissue were taken into account. The major acids in animal fats and in fish oils are myristic, palmitic, palmitoleic, stearic, oleic, eicosenoic, arachidonic, EPA, docosenoic and DHA.

1.3   Saturated acids (Table 1.1)

1.3.1   Short- and medium-chain saturated acids (C_4–C_{14})

Members of this group of acids occur in milk fats and in some vegetable oils known as the lauric oils. Cow milk fat, for example, contains butanoic (butyric, C_4) acid at a level of about 4% (wt). This may not seem very much but because of the low molecular weight of the C_4 acid compared

Table 1.1 Names and selected physical properties of some alkanoic acids

Chain length	Systematic name	Trivial name	Acid		Methyl ester		Mol wt
			mp (°C)	bp (°C)[a]	mp (°C)	bp (°C)[a]	
4	butanoic	butyric	−5.3	164	–	103	88.1
6	hexanoic	caproic	−3.2	206	−69.6	151	116.2
8	octanoic	caprylic	16.5	240	−36.7	195	144.2
10	decanoic	capric	31.6	271	−12.8	228	172.3
12	dodecanoic	lauric	44.8	130[1]	5.1	262	200.3
14	tetradecanoic	myristic	54.4	149[1]	19.1	114[1]	228.4
16	hexadecanoic	palmitic	62.9	167[1]	30.7	136[1]	256.4
18	octadecanoic	stearic	70.1	184[1]	37.8	156[1]	284.5
20	eicosanoic	arachidic	76.1	204[1]	46.4	188[2]	312.5
22	docosanoic	behenic	80.0	–	51.8	206[2]	340.6
24	tetracosanoic	lignoceric	84.2	–	57.4	222[2]	368.6

[a]bp at 760 mm or at 1 or 2 mm as indicated by superscript.
Adapted from *The Lipid Handbook*, 2nd edition (1994) p. 1.

with the more common C_{18} acids this represents about 8.5% on a molar basis, which means that it could be present in up to 25% of milk fat glycerides. Also present are lower levels of the C_6–C_{12} acids. The short-chain acids are retained in butter and in other products made from milk fat.

Some seed oils contain very high levels of lauric acid (C_{12}, about 50%) with significant levels of caprylic (C_8), capric (C_{10}) and myristic acids (C_{14}) also. The best known of these oils are coconut oil and palm kernel oil. At the present time there is interest in other sources of these medium-chain acids (section 3.7.1). Among these are the *cuphea* oils which show great promise but still need to be fully domesticated before they produce an easily cultivated crop. A transgenic rapeseed has also been developed which produces a lauric-rich oil.

1.3.2 Palmitic and stearic acids

Palmitic acid (16:0) is the most widely occurring saturated acid. It is present in fish oils (10–30%), in milk and body fats of land animals (up to 30%), and in virtually all vegetable fats at a range of levels between 5 and 50%. Useful sources of palmitic acid include cottonseed oil (15–30%), palm oil (30–60%), Chinese vegetable tallow (60–70%), lard (20–30%), and tallow from sheep and cattle (25–35%).

Despite popular impression, stearic acid (18:0) is much less common than palmitic acid. It is a major component of the tallows of ruminant animals (5–40%) and a significant component in a number of vegetable tallows (solid fats of vegetable origin) including cocoa butter (30–35%), Illipe or Borneo tallow (40%), and shea butter (45%). Stearic acid is also easily made by hydrogenation of readily available oleic, inoleic and linolenic acid.

Since many population groups in the world have scruples about animal products vegetable sources of palmitic acid and stearic acid are preferred. These are used in food and non-food products (surfactants, cosmetics, personal hygiene products) to obtain products which find general acceptance.

1.3.3 Long-chain acids

Saturated acids of chain length greater than 18 carbon atoms are present at low levels in a few seed oils and at higher levels only in a few uncommon sources. The C_{20}–C_{30} members are often present in waxes. A convenient source of some of these acids is groundnut oil which contains 5–8% of long-chain acids including arachidic (20:0), behenic (22:0), and lignoceric (24:0) acids. Rarer but richer sources include rambutan tallow (*Nephelium lappaceum*, about 35% of 20:0), kusum (*Schleichera trifuga*, 20–30% of 20:0), *Lophira alata* (15–30% of 22:0), and *L. procera* (around 20% of 22:0) seed fats. Long-chain acids can also be made from more readily available shorter chain acids (C_{12}–C_{18}) by appropriate chain-extension procedures. These reaction sequences are not confined to adding one or two carbon atoms at a time since methods exist for adding five or six carbon atoms in one cycle of reactions.

1.4 Monoene acids (Table 1.2)

Over 100 monoene acids have been described. These fall almost entirely in the range C_{10}–C_{30} with C_{16}, C_{18}, and C_{22} members the most common. Most have *cis* (*Z*) configuration and the most important are either $\Delta 9$ or *n*-9, i.e. the double bond is nine carbon atoms from the carboxyl ($\Delta 9$) or the methyl group (*n*-9).

9-hexadecenoic acid (palmitoleic, zoomaric) is a minor component (<1%) of many seed oils and animal fats. It is more significant in fish oils where it may attain a level around 10% and is a major component of a few seed oils. Macadamia oil is a useful source of this acid at around the 20% level. It is also a significant component in *Saccaromyces cerevisiae* (30–50%) and other yeasts.

9-octadecenoic acid (oleic) is the most widely distributed and the most extensively produced of all fatty acids (see section 1.2). It is the prototype of all monoene acids and serves as the biological precurser of other *n*-9 monoene acids and of the *n*-9 family of polyene acids. Olive oil (60–80%) and almond oil (60–70%) are rich sources of this acid and it is also present at high levels (55–75%) in several nut oils (filbert, cashew, pistachio,

Table 1.2 The more common *cis*-monoene acids

Systematic name	Trivial name	CH$_3$(CH$_2$)$_n$CH=CH(CH$_2$)$_m$COOH	
		n	*m*
9-hexadecenoic	palmitoleicic[b]	5	7
9-octadecenoic	oleic	7	7
9-octadecenoic[a]	elaidic	7	7
6-octadecenoic	petroselinic	10	4
11-octadecenoic	*cis*-vaccenic[c]	5	9
11-octadecenoic[a]	*trans*-vaccenic	5	9
9-eicosenoic	gadoleic	9	7
11-eicosenoic	gondoic	7	9
5-docosenoic	–	15	3
11-docosenoic	cetoleic	9	9
13-docosenoic	erucic	7	11
15-tetracosenoic	nervonic[d]	7	13

[a]*trans* isomer
[b]also zoomaric
[c]also ascelepic
[d]also selacholeic

pecan, macadamia – see Table 3.8). It is a significant component in low-erucic rapeseed oil, groundnut oil, and palm oil. High-oleic varieties of sunflower oil (80% and above) and safflower oil (70–75%) are now available and the former is commonly used as a source of oleic-rich glycerides. Oleic acid is a major component in most animal fats where it is often accompanied by low levels of other 18:1 isomers. This may or may not matter depending on the end-use of the acid.

Other octadecenoic acids include petroselinic (6*c*), elaidic (9*t*), and vaccenic (11*c* and 11*t*). The *trans*-isomers are discussed later and *cis*-vaccenic acid is not important. It is an *n*-7 acid produced by chain elongation of 9-hexadecenoic acid. Petroselinic acid is a major component (35–85%) of many seed oils of the families *Umbelliferae*, *Araliaceae*, and *Garryaceae*. Examples include carrot, caraway, parsnip, and parsley. Attempts are being made to develop coriander as a new crop containing about 80% of petroselinic acid.

Eicosenoic acid present in fish oils is often a mixture of the Δ9 (*n*-11) and Δ11 (*n*-9) isomers. The Δ11 acid frequently accompanies erucic acid in oils rich in this acid.

The most important docosenoic acid (22:1) is erucic acid (Δ13*c*, *n*-9) present at high levels in seed oils of the Cruciferae. This acid has been bred out of rapeseed oil grown for food use but there is a considerable demand for erucic acid in the form of its amide (sections 8.7.1 and 10.6). Erucic acid is available on a commercial scale from high-erucic rapeseed oil (45–50%), mustard seed oil (around 60%), and from the seed oil of *Crambe* *abyssinica* (about 60%). In fish oils the major docosenoic acid is the Δ11 isomer (cetoleic).

Jojoba oil (from *Simmondsia chinensis*) is unusual in that its major lipids are ester waxes (esters of long-chain alcohols and long-chain acids) rather than glycerol esters. Most of the acids and alcohols are C_{20} and C_{22} compounds with one double bond. Meadowfoam oil (*Limnanthes alba*) is unusual in that more than 95% of its fatty acids are C_{20} and C_{22} compounds. These are mainly 20:1 (5*c*, 63%) and 22:1 (5*c*, 3% and 13*c*, 10%) along with 22:2 (5*c*13*c*, 18%).

The major C_{24} monoene acid is nervonic (Δ15*c*, *n*-9). This is a significant component of many sphingolipids. Honesty seed oil (*Lunaria biennis*), with C_{22} (43%) and C_{24} (25%) monoene acids, is probably the most convenient source of nervonic acid. This acid is being investigated for the treatment of multiple sclerosis.

1.5   Methylene-interrupted polyene acids

The most important polyene acids have a methylene-interrupted pattern of unsaturation with 2–6 double bonds and *cis* (*Z*) configuration. They are grouped in families and are biosynthetically related to other members of the same family. The two major groups are the *n*-6 acids based on linoleic acid and the *n*-3 acids based on α-linolenic acid. Two minor groups are based on oleic acid (*n*-9 family) and on 9-hexadecenoic acid (*n*-7 family). The common and important C_{18} members are present in most vegetable oils while the C_{16}, C_{20} and C_{22} acids are present in fish oils and in the lipids of other animals. Some of these acids cannot be made by animals and are essential dietary requirements of vegetable origin (see section 1.15.4).

Table 1.3 Methylene-interrupted polyene acids

Trivial name	Structure	Double bond position[a]
linoleic	18:2 (*n*-6)	9,12
γ-linolenic	18:3 (*n*-6)	6,9,12
α-linolenic	18:3 (*n*-3)	9,12,15
stearidonic	18:4 (*n*-3)	6,9,12,15
dihomo-γ-linolenic	20:3 (*n*-6)	8,11,14
Mead's acid	20:3 (*n*-9)	5,8,11
arachidonic	20:4 (*n*-6)	5,8,11,14
eicosapentaenoic (EPA)	20:5 (*n*-3)	5,8,11,14,17
docosapentaenoic (DPA)	22:5 (*n*-3)	7,10,13,16,19
docosahexaenoic (DMA)	22:6 (*n*-3)	4,7,10,13,16,19

[a]All double bonds have *cis* configuration.

Structures and names are given in Table 1.3 and some comments on the more important *n*-6 and *n*-3 acids follow.

Linoleic acid (18:2, *n*-6) is the most common of all the polyene acids and serves as a prototype for other acids in this category. It is present in all vegetable fats and is a major component in several. Among the more common sources are soybean (45–60%), low-erucic rapeseed (10–30%), groundnut (15–45%), rice bran (35–40%), poppy (around 70%), safflower (55–80%), sunflower (20–75%), corn (maize) (35–60%), cottonseed (35–60%), sesame (about 45%), and tall oil fatty acids (40–50%).

γ-linolenic acid (18:3, *n*-6, GLA) is not a common acid but recent interest in its use as a dietary supplement has directed attention to its three common sources: evening primrose oil (8–12%), borage oil (20–25%), and blackcurrant seed oil (15–17%) (see section 3.2.1).

Arachidonic acid (20:4, *n*-6, AA) is of considerable importance as the precursor of many important C_{20} metabolites such as the prostaglandins, thromboxanes, and leukotrienes (members of the eicosanoid cascade). This acid is a minor component of many fish oils and attains higher levels in animal phospholipids such as those from egg and from liver. Though rare in the plant kingdom, it has been identified in mosses, ferns, and some algae and fungi.

α-linolenic acid (18:3, *n*-3) is an important component of lipids in leaves, stems, and roots. It is a minor but significant component of soybean oil (4–11%) and of rapeseed oil (5–15%) and a major component of linseed oil (50–60%) and perilla oil (60–70%). EPA (20:5) and DHA (22:6) are also important *n*-3 acids. The former is a precursor of a series of eicosanoids and the latter is present at high levels in the lipids of sperm and the retina. The major source of these two acids are the fish oils which may contain up to 30% of the two acids combined. One or both are also produced by some micro-organisms.

Mead's acid (20:3, *n*-9) is of interest in that it becomes significant in animal lipids only during periods of essential fatty acid deficiency (section 1.15.4).

1.6 Conjugated polyene acids

This section is confined to polyolefinic compounds and excludes acids having acetylenic unsaturation and/or oxygenated functions. Such acids are discussed later. The best known conjugated polyene acids have 18 carbon atoms and three or four double bonds per molecule. Two Δ9,11,13,15 tetraenes are derived biologically from α-linolenic acid (Δ9,12,15), and Δ9,11,13 and Δ8,10,12-trienes are derived from linoleic acid (Δ9,12). These are listed in Table 1.4 with structures, trivial names, and major sources.

Table 1.4 Natural octadecatrienoic and octadecatetraenoic acids with conjugated unsaturation

Configuration	Trivial name	Typical source (seed oils)
trienes (18:3)		
8c10t12c	jacaric	*Jacaranda mimosifolia*
8t10t12c	calendic	*Calendula officinalis*
8t10t12t	–	*C. officinalis*
9c11t13c	catalpic	*Catalpa ovata*
9c11t13t	α-eleostearic[a,b]	*Aleurites fordii*[d]
9t11t13t	β-eleostearic	*A. fordii*[d]
9t11t13c	punicic	*Punica granatum*
tetraenes (18:4)		
9c11t13t15c	α-parinaric[c]	*Impatiens balsamina*
9t11t13t15t	β-parinaric	–

[a]also kamlolenic 18-hydroxy-α-eleostearic (*Mallotus philippinensis*)
[b]also licanic 4-oxo-α-eleostearic (*Licania rigida*)
[c]also chrysobalanic 4-oxo-α-parinaric (*Chrysobalanus icaco*)
[d]tung oil
Adapted from *The Lipid Handbook*, 2nd edition (1994) p. 7.

1.7 Other polyene acids

Some polyene acids have incompletely methylene-interrupted unsaturation. These are found in some seed oils, some mycobacteria, and some marine lipids, especially sponges. The additional double bond is frequently $\Delta 3$ or $\Delta 5$. Examples include a diene acid (20:2 5c13c) in meadowfoam oil, pinolenic acid (18:3 5c 9c12c) in tall oil, columbinic acid (18:3 5t 9c12c) in aquilegia oil (columbine) and long-chain trienes such as 26:3 (5,9,19), 28:3 (5,9,19), and 30:3 (5,9,23) in sponge lipids (demospongic acids).

1.8 Acetylenic and allenic acids

Acetylenic acids occur only rarely in seed oils and in mosses. Structures have been identified containing acetylenic unsaturation only (e.g. tariric), mixed olefinic and acetylenic unsaturation which may be conjugated (ximenynic) or non-conjugated (crepenynic), and yet others which also contain a hydroxyl group (helenynolic).

A few allenic acids are also known and are probably produced by rearrangement of acetylenic precursors. The allenic acids are optically active by virtue of the allenic group which is a chiral unit. Some of these structures are listed in Table 1.5.

Table 1.5 Acetylenic and allenic acids[a]

Trivial name	Structure	
non-conjugated		
tariric	18:1 6a[b]	
stearolic	18:1 9a	
crepenynic	18:2 9c12a	
dehydrocrepenynic	18:3 9c12a14c	
–	18:3 6a9c12c	
–	18:4 6a9c12c15c	
–	20:3 8a11c14c	
conjugated		
pyrulic	17:2 8a10t	
ximenynic (santalbic)	18:2 9a11t	
isanic (erythrogenic)	18:3 9a11a17e	
bolekic	18:3 9a11a13c	
exocarpic	18:3 9a11a13t	
allenic		
laballenic	18:2 5e6e	R(−)
lamenallenic	18:3 5e6e16t	R(−)

[a]Many of these acids also occur as hydroxy derivatives (see section 1.12.1).
[b]These numbers indicate the total number of unsaturated centres independent of their nature.
R indicates absolute configuration.
− indicates sign of rotation.

1.9 Acids with *trans*-olefinic unsaturation

Polyene acids having *trans*-olefinic groups in conjugated and non-methylene-interrupted systems have already been discussed. Natural acids with *trans* unsaturation are very uncommon. Nevertheless they are a significant component (about 5%) of human dietary intake. They arise from two major sources: first from some cooking fats and fat-based spreads which have been produced by partial hydrogenation with a heterogeneous catalyst (section 4.3), and secondly from dairy fats and other fats from ruminants. The latter result from biohydrogenation of linoleic and linolenic acid in the animal's diet. For example, butter fat contains ten 18:1 *trans*-acids of which about 75% is *trans*-vaccenic acid (Δ11t). The *trans*-acids in human milk fat (2–4%) probably arise from these dietary sources. Yet a third source of dietary *trans*-acids is their presence in some refined oils caused by exposure to high temperatures (230–250°C) during the refining process. The major acids of this type, present for example in refined soybean and rapeseed oil, are the 9c12c15t and 9t12c15c isomers from linolenic acid. Linoleic acid undergoes a similar reaction, but less readily, to give the 9c12t and 9t12c diene acids (section 4.2).

1.10   Branched-chain acids

There are many branched-chain acids although these seldom occur at a significant level. Examples include the iso acids (mainly with an even number of carbon atoms) and the anteiso acids (mainly with an odd number of carbon atoms) found in wool wax and other animal fats, polymethyl branched acids in bacterial lipids, and phytol-based acids present at low levels in fish oils.

$$CH_3$$
$$|$$
$$CH_3CH(CH_2)_nCOOH$$
iso acids

$$CH_3$$
$$|$$
$$CH_3CH_2CH(CH_2)_nCOOH$$
anteiso acids

$$\begin{array}{ccc} CH_3 & CH_3 & CH_3 \\ | & | & | \end{array}$$
$$CH_3(CH_2)_nCHCH_2CHCH_2CHCOOH$$
2,4,6-trimethylalkanoic acids   (e.g. C_{25}, C_{27}, and C_{29}, acids)

$$\begin{array}{cccc} CH_3 & CH_3 & CH_3 & CH_3 \\ | & | & | & | \end{array}$$
$$CH_3(CH_2)_nCHCH_2CHCH_2CHCH_2CHCOOH$$
2,4,6,8-tetramethylalkanoic acids   (e.g. C_{30}, C_{32}, C_{34} acids)

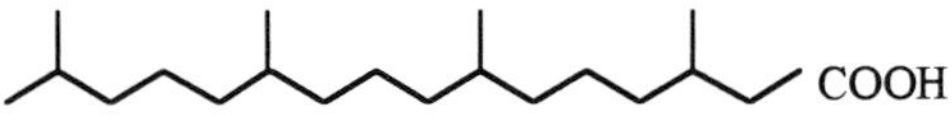

COOH   phytanic acid (C_{20})

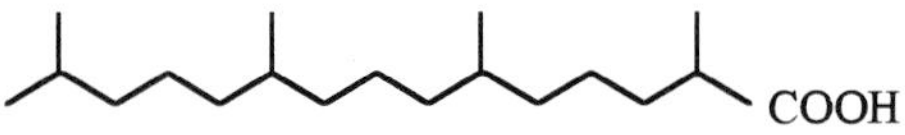

COOH   pristanic acid (C_{19})

1.11   Cyclic acids

The most common cyclic acids contain a cyclopropane, cyclopropene, or cyclopentene unit. Cyclopropane acids occur in bacterial membrane phospholipids and are mainly C_{17} or C_{19} (lactobacillic) acids. The cyclopropane unit, like a *cis* double bond, introduces a discontinuity in the molecule and increases fluidity in the membrane.

$$CH_2$$
$$\bigwedge$$
$$CH_3(CH_2)_5CHCH(CH_2)_nCOOH$$
$$n = 7 \text{ or } 9$$

The best known cyclopropene acids are malvalic and sterculic which are present at high levels in sterculia oils and at lower levels in kapok seed oil (around 12%) and in cottonseed oil (about 1%). These acids are highly reactive and are destroyed during refining and hydrogenation of the oils. They have attracted interest because they inhibit the biodesaturation of stearic to oleic acid. 2-hydroxysterculic acid has also been identified. This compound is probably an intermediate in the bioconversion of sterculic to malvalic acids by α-oxidation (section 7.4.1).

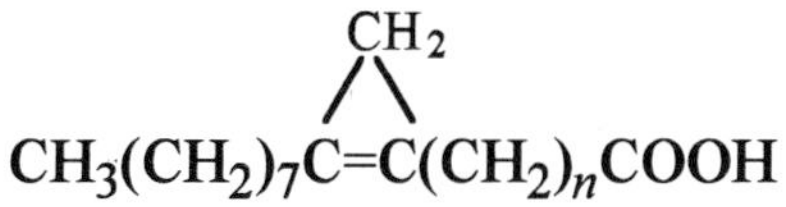

$$CH_2$$

$$CH_3(CH_2)_7C=C(CH_2)_nCOOH$$

malvalic acid $n = 6$

sterculic acid $n = 7$

Seed fats of the *Flacourtiaceae* are unique in containing several cyclopentene acids. They range from C_6 to C_{20} acids and may also be unsaturated in the side chain (Table 1.6). These acids are optically active by virtue of the chiral centre in the cyclopentene unit.

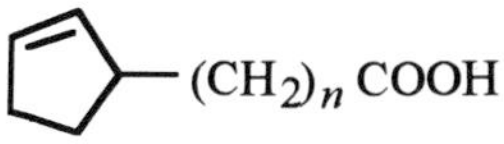

$$(CH_2)_n COOH$$

Cyclic acids are formed in small amounts from polyene acids, especially linolenate, when exposed to high temperatures during refining or when used for frying (section 4.2).

1.12 Oxygenated acids

The most common of the oxygenated acids contain hydroxy, epoxy, or furanoid units. Other systems not included here include keto(oxo) and methoxy groups.

1.12.1 Hydroxy acids

The best known natural hydroxy acid is ricinoleic acid (12-hydroxyoleic). This is the major acid (around 90%) in castor oil (*Ricinus communis* seed oil). An isomer (9-OH 18:1 12c) occurs in *Strophanthus* and *Wightia* seed oils. Other acids of this type include densipolic, lesquerolic, and auricolic (Table 1.7). Lesquerolic acid is the C_{20} homologue of ricinoleic acid and attempts are being made to develop *Lesquerella fondleri* as a new crop.

Table 1.6 Natural cyclopentenyl acids

Trivial name	Number of carbon atoms	n^a	x^a	y^a	Configuration
aleprolic	6	0	–	–	$R(+)$
alepramic	8	2	–	–	$S(+)$
aleprestic	10	4	–	–	$R(+)$
aleprylic	12	6	–	–	$R(+)$
alepric	14	8	–	–	$R(+)$
hydnocarpic	16	10	–	–	$R(+)$
chaulmoogric	18	12	–	–	$R(+)$
hormelic	20	14	–	–	$R(+)$
–	16	–	2	6	–
manaoic	16	–	4	4	–
–	16	–	7	1	–
gorlic	18	–	4	6	$R(+)$
–	18	–	7	3	–
oncobic	20	–	6	6	–
–	20	–	7	5	

[a]Values of n, x, and y based on the following structures

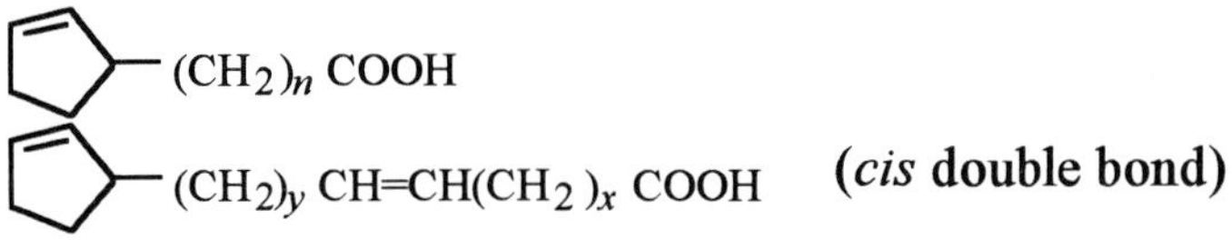

$$\square\!\!\!\!\diagdown\!-(CH_2)_n\ COOH$$

$$\square\!\!\!\!\diagdown\!-(CH_2)_y\ CH{=}CH(CH_2)_x\ COOH \qquad (\textit{cis } \textbf{double bond})$$

Adapted from *The Lipid Handbook*, 2nd edition (1994) p. 13.

Table 1.7 Mid-chain hydroxy acids without conjugated unsaturation

Chain length and unsaturation	Position of hydroxyl group	Trivial name	Source (seed oils)
16:1 9c	12	–	*Lesquerella densipila*
18:1 9c	12R	ricinoleic	*Ricinus communis*
18:1 12c	9S	isoricinoleic	*Strophanthus* and *Wightia* spp.
18:2 9c15c	12R	densipolic	*L. densipila*
20:1 11c	14R	lesquerolic	*L. densipila*
20:2 11c17c	14	auricolic	*L. auriculata*

Adapted from *The Lipid Handbook*, 2nd edition (1994) p. 15.

Castor oil and ricinoleic acid are important materials used in cosmetics, in lubricants both before and after hydrogenation, and as a drying oil after dehydration (dehydrated castor oil or DCO). Also important are the products of pyrolysis (heptanal and 10-undecenoic acid and materials derived from these) and of alkali fusion (2-octanol, 2-octanone, 10-hydroxydecanoic acid and the C_{10} dibasic acid, sebacic). Of less importance are some hydroxy acids based on 8-, 9-, or 13-hydroxystearic acid with conjugated unsaturation. Some typical examples are:

isanolic   8-OH 18:3 9a,11a,17e
helenynolic   9-OH 18:2 10t,12a
dimorphecolic   9-OH 18:2 10t,12c
coriolic   13-OH 18:2 9c,11t

1.12.2 Epoxy acids

Several natural epoxy acids are known (Table 1.8). Vernolic was first discovered in 1954 and is still the best known in this group. It occurs at high levels in the seed oils of *Vernonia anthelmintica* (70–75%), *V. galamensis* (73–78%), *Cephalocroton cordofanus* (60–65%), *Euphorbia lagascae* (60–65%), *Stokes aster* (65–80%), and *Erlanga tomentosa* (50–55%), and attempts are being made to develop *V. galamensis* and *E. lagascae* as commercial crops.

On prolonged storage of seeds, unsaturated acids may become oxidized and 9,10-epoxystearate and 9,10-epoxy-12-octadecenoate have been identified. These are formed in an optically active form and result, presumably, from oleate and linoleate by enzymic oxidation. These and other epoxy acids are listed in Table 1.8.

1.12.3 Furanoid acids

Natural furanoid acids with the structures shown below have been identified. They are present at low levels, often as a complete series, in fish

Table 1.8 Epoxy acids in seed lipids

Chain length and unsaturation	Position of epoxide group[a]	Trivial name
18:0	9,10	–
18:1 12c	9,10 (9R,10S)	coronaric
18:1 12a	9,10	–
18:1 9c	12,13 (+) 12S,13R	vernolic
18:1 9c	12,13 (−) 12R,13S	vernolic
18:2 12c15c	9,10	–
18:2 6c9c	12,13	–
18:2 9c12c	15,16	–
18:2 3t12c	9,10	–
18:2 6t9c	12,13	–
20:1 11c	14,15	alchornoic

[a]All the epoxide functions have *cis* configuration except 9,10-epoxystearic acid which exists in *cis* and *trans* forms

oils. The proportion of these increases significantly during prolonged fasting. The short-chain urofuranic acids have been identified in animal blood and urine. They are probably breakdown products of furanoid acids which are believed to come from dietary vegetable sources.

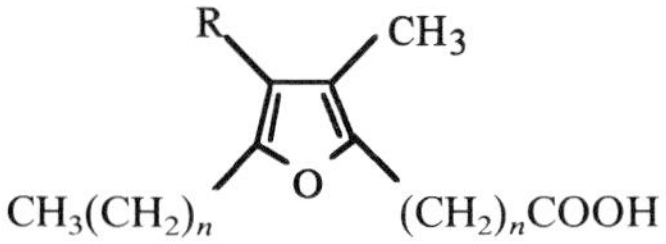

$R = CH_3$ or H

$m = 2$ or 4 $n = 6,8,$ or 10

furanoid acids

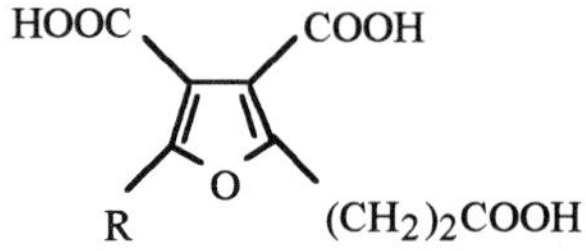

$R = CH_3(CH_2)_4$ or $CH_3(CH_2)_2$

or CH_3CH_2CHOH or CH_3CH_2CO

urofuranic acids

1.13   Other acids

In addition to the major and minor fatty acids which have been surveyed in this chapter there are still others which include unusual features. 16-fluoropalmitic and 18-fluoro-oleic acids are present in seeds of *Dichapetalum toxicarium* and are responsible for its high toxicity. Other halogenated acids occur mainly in marine sources where chlorine, bromine, and iodine are all available and three C_{19} sulphur-containing acids have been identified in rapeseed oil. The following structures are examples:

$CH_3(CH_2)_nCH=CHCH_2CH_2CBr=CH(CH_2)_3COOH$
$n = 13$ and 14

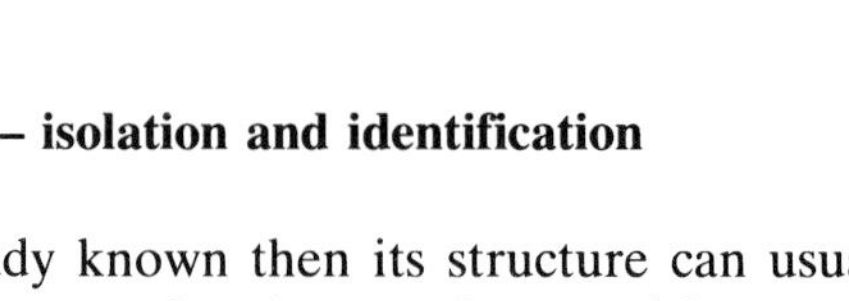

$CH_3(CH_2)_x$ $(CH_2)_{12-x}COOH$

$x = 5,6,$ or 7

1.14   Fatty acids – isolation and identification

If an acid is already known then its structure can usually be defined by comparison with an authentic sample or with compounds of related structure. This comparison is generally made on the basis of its gas chromatographic behaviour (i.e. its retention time on one or more columns) but may include other chromatographic and/or some spectroscopic investigations. If the acid has a completely novel structure then further investigation will be required and spectroscopic procedures are likely to provide the major evidence (sections 6.2–6.7).

Evidence that an unusual acid is present may be apparent from unexpected behaviour during routine chromatographic or spectroscopic examination. It is then desirable to isolate or concentrate the unusual acid prior to its identification in an appropriate manner. Chromatographic procedures of separation are now often combined with spectroscopic methods of identification in a single operation. These are described as hyphenated techniques and include gas chromatography-mass spectrometry (GC-MS), liquid chromatography-gas chromatography (LC-GC), liquid chromatography-Fourier transform infrared spectroscopy (LC-FTIR), etc.

To fully define the structure of a fatty acid it is necessary to know:

1. its chain length and whether the molecule contains any branched or cyclic systems or any of the less common functional groups;
2. the number, nature, configuration, and position of all unsaturated centres;
3. the nature, position, and (where relevant) the stereochemical configuration of any functional groups.

This discussion is confined to those separation procedures which are exploited for analytical purposes.

The classical separation procedures of distillation and crystallization are only of limited value to lipid chemists. With polyunsaturated acids it is desirable to avoid the high temperatures associated with fractional distillation since these promote undesirable changes such as stereomutation, double bond migration, cyclization, and dimerization. Distillation permits separation by chain length and is used industrially for this purpose. It is of only limited value for separating acids/esters differing only in their degree of unsaturation. Limited use is still made of urea fractionation.

Urea normally crystallizes in tetragonal form, but in the presence of certain aliphatic compounds it forms hexagonal prisms containing some of the aliphatic material. These prisms are built up from urea: six molecules form a unit cell 11.1×10^{-10} m long and 8.2×10^{-10} m in diameter containing a channel in which an open-chain molecule may be held so long as it fulfills certain dimensional qualifications. It must not be too short or it will not be held within the channel, and it must not be too wide if it is to fit into the free space, variously estimated at between 4.0 and 6.0×10^{-10} m. Many straight-chain acids and their esters satisfy these conditions and thus readily form complexes (adducts, inclusion compounds) with urea.

Saturated acids form stable complexes more readily than do the unsaturated acids, and oleic acid enters into these inclusion compounds more readily than do the polyunsaturated acids. In practice urea and mixed acids are dissolved in hot methanol or urea and methyl esters in a hot methanol–ethanol mixture. The solution is crystallized at room temperature or at 0°C. The adduct and mother liquor will furnish the acids or esters when mixed with water and extracted with ether or petroleum ether in the usual way.

This procedure is employed for two purposes. It separates straight-chain acids or esters from branched-chain or cyclic compounds with the former concentrating in the adduct and the latter in the mother liquor. Urea fractionation is also used to separate acids or esters of differing unsaturation.

Many fatty acids differ so little in solubility that they can rarely be purified by crystallization. Most useful separations are now achieved by chromatography some of which can generally provide sufficient material for subsequent spectroscopic examination.

1.14.1   Thin-layer chromatography

Thin-layer chromatography is still a useful isolation or concentration process. Simple adsorption on silica will not usefully separate acids or esters differing only in chain length and/or extent of unsaturation. Nevertheless there is some subfractionation according to these factors. Long-chain compounds are slightly less polar than their medium-chain analogues and polyene compounds are slightly more polar than their saturated homologues. Acids or esters with additional polar functional groups (hydroxy, epoxy, etc.) are well separated depending on the number of such groups, and thin-layer chromatography provides an easy way of isolating and separating such compounds.

Other separations can be achieved if the usual silica layer is modified in some way. The best known example of this is silver ion chromatography. Silver nitrate at a level between 5 and 20% is incorporated into the silica and acids/esters are then separated mainly according to the number of double bonds (Figure 1.1). Stereoisomers (cis and trans) can also be separated from each other. If a sample contains only cis-olefinic unsaturation then it is possible to separate saturated monoene, diene, triene, etc. compounds (section 5.2.4c).

A recently described method employs Bond-elut columns charged with silver nitrate–acetonitrile–water. Esters (around 0.5 mg) of differing unsaturation are placed on the column and eluted with appropriate solvents (Table 1.9). Though small, these fractions are large enough for further examination by gas chromatography.

Borates and arsenites have also been incorporated into silica layers. These form complexes with polyhydroxy compounds and lead to separations based on the number of hydroxyl groups, their position in the carbon chain, and their stereochemistry.

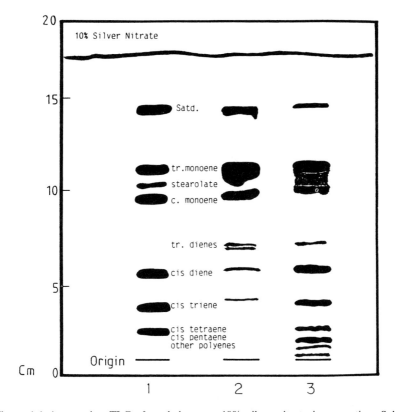

Figure 1.1 Argentation TLC of methyl esters. 10% silver nitrate impregnation. Solvent diethyl ether/hexane 8:92. Lane 1 = standards of unsaturation as shown, including methyl stearolate (methyl 9-octadecynoate). Lane 2 = *trans*-hardened soybean oil. Lane 3 = blended hardened and unhardened fish oils; note the confused area between the *cis* and *trans* monoenes. Reproduced with permission (Hammond, 1993).

Table 1.9 Solvents used to elute methyl esters of differing unsaturation from a Bond-elut silver ion column

Esters	Volume (ml)	Solvent		
		CH_2Cl_2 (%)	$COMe_2$ (%)	CH_3CN (%)
saturated	5	100	–	–
monoene	5	90	10	–
diene	5	–	100	–
triene	10	–	97	3
tetraene	10	–	94	6
pentaene	5	–	88	12
hexaene	5	–	60	40

1.14.2   High-performance liquid chromatography (HPLC)

Separations effected by thin-layer chromatography can usually be achieved more efficiently on columns using HPLC systems. This can also be extended to silver ion systems and can be carried out on a semi-preparative scale.

1.14.3   Gas chromatography

Gas chromatography is now the technique most commonly employed to separate methyl esters for quantitative analytical purposes. Originally packed columns were used but the more efficient capillary columns are now commonly employed (section 5.2.3). The separated components are not generally isolated but a careful study of elution behaviour may indicate chain length, degree of unsaturation, and possibly the position of unsaturated centres. For identification purposes there are some advantages in operating under isothermal conditions. Retention times can be compared with those of an internal standard which is usually a saturated ester such as palmitate or stearate, or the retention behaviour can be expressed in terms of equivalent chain length (ECL) first suggested by Miwa *et al.* in 1960. This is the notional number of carbon atoms in a saturated ester whose methyl ester would co-elute with the ester in

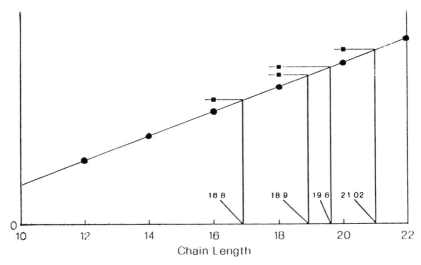

Figure 1.2 Plot of the log of retention time against chain length for a series of methyl esters on a polar phase packed GLC column. Note that there is a straight line relationship and that double bonds effectively increase the equivalent chain length of the ester. Column was 2 m × 4 mm i.d. glass, packed with Supelco SP2330 phase. ——●—— monoenoic FAME; ——■—— dienoic FAME. Reproduced with permission (Hammond, 1993).

Table 1.10 Gas chromatographic equivalent chain lengths of methyl esters of some natural fatty acids on different stationary phases

Fatty acid	Stationary phase			
	Silicone	Carbowax	Silar 5CP	CP-Sil 84
14-iso	14.64	14.52	14.52	14.51
14-anteiso	14.71	14.68	14.68	14.70
14:1 (n-5)	13.88	14.37	14.49	14.72
16-iso	16.65	16.51	16.51	16.50
16-anteiso	16.73	16.68	16.68	16.69
16:1 (n-7)	15.83	16.25	16.38	16.60
16:2 (n-4)	15.83	16.78	16.98	17.47
16:3 (n-3)	15.69	17.09	17.31	18.06
16:4 (n-3)	15.64	17.62	17.77	18.82
17:1 (n-8)	16.75	17.19	17.33	17.51
18:1 (n-9)	17.73	18.16	18.30	18.47
18:1 (n-7)	17.78	18.23	18.36	18.54
18:2 (n-6)	17.65	18.58	18.80	19.20
18:3 (n-6)	17.49	18.85	19.30	19.72
18:3 (n-3)	17.72	19.18	19.41	20.07
18:4 (n-3)	17.55	19.45	19.68	20.59
20:1 (n-11)	19.67	20.08	20.22	20.35
20:1 (n-9)	19.71	20.14	20.27	20.41
20:3 (n-9)	19.24	20.66	20.92	21.43
20:3 (n-6)	19.43	20.78	21.05	21.61
20:3 (n-3)	19.71	20.95	21.22	21.97
20:4 (n-6)	19.23	20.96	21.19	21.94
20:4 (n-3)	19.47	21.37	21.64	22.45
20:5 (n-3)	19.27	21.55	21.80	22.80
22:1 (n-11)	21.61	22.04	22.16	22.30
22:1 (n-9)	21.66	22.11	22.23	22.36
22:4 (n-6)	21.14	22.90	23.21	23.90
22:5 (n-6)	20.99	23.15	23.35	24.19
22:5 (n-3)	21.18	23.50	23.92	24.75
22:6 (n-3)	21.04	23.74	24.07	25.07

n-saturated acids are excluded from this table. By definition the ECL is equal to the number of carbon atoms in the alkanoic acid.
Adapted with permission from *Gas Chromatography and Lipids* (1989) W.W. Christie, The Oily Press, Dundee, p. 104.

question. For example, if methyl oleate has an ECL of 18.50 this means that it would elute between the C_{18} and C_{19} methyl esters at the same time as a saturated ester with 18½ carbon atoms. This concept is based on the observation that for any homologous series eluted under isothermal conditions the plot of log retention time against the number of carbon atoms is linear. Some examples are given in Figure 1.2 and Table 1.10.

Among 18:3 isomers, for example, the order of elution on a polar column is given by $\Delta 5,9,12$ before $\Delta 6,9,12$ before $\Delta 9,12,15$.

1.14.4   Spectroscopic procedures of identification

The identification of a novel acid would, nowadays, lean very heavily on spectroscopic procedures including ultraviolet, infrared, Raman, ^{1}H and ^{13}C nuclear magnetic resonance (NMR), and mass spectrometry. These topics are covered in sections 6.2–6.7. Hyphenated techniques such as GC-MS are commonly employed. This combines the separating power of gas chromatography with the ability to identify structure which is implicit in mass spectrometry.

1.14.5   Chemical procedures of identification

With the increasing use of spectroscopic techniques to determine structure, less use is made of the chemical procedures which were so important at one time. However, two are still worthy of discussion: hydrogenation and oxidative cleavage.

Complete catalytic hydrogenation of an unsaturated acid/ester using a nickel, platinum, or palladium catalyst followed by identification of the perhydro compound – probably by gas chromatography – will show whether the carbon chain is straight or branched chain or contains a cyclic unit and also its total length. Many unsaturated acids are reduced to stearic acid showing that they are straight-chain C_{18} compounds.

Sometimes there is an advantage in carrying out partial hydrogenation. This is effected with hydrazine which, in contrast to reduction with heterogeneous catalysts, occurs without double bond migration and without stereomutation. Double bonds remaining after partial hydrogenation have the same position and the same configuration as in the original polyene acid. This approach is particularly useful for compounds containing *cis* and *trans* unsaturation and an example is described in the next paragraph.

Partial hydrogenation of 18:3 (9t12c15c) will give three dienes (9t12c, 12c15c, and 9t15c), three monoenes (9t,12c, and 15c), and stearate. After separation of the *cis*- and *trans*-monoenes it can then be shown by oxidation (see below) that the *trans*-monoene is the 9t isomer and the *cis*-monoenes are a mixture of 12c and 15c isomers. These together confirm the structure of the original triene.

Oxidation of unsaturated acids or esters or alcohols is usually achieved by reaction with ozone (ozonation) followed by cleavage of the ozonide under conditions summarized in Table 1.11. Further details are available in section 7.5.3. Monoene compounds give one monofunctional product and one bifunctional product and these can usually be identified by gas

Table 1.11 Decomposition of ozonides – products and reagents

Products	Reagents
alcohols	$LiAlH_4$, $NaBH_4$, H_2/Pd-C
aldehydes	Zn/H^+, Ph_3P, Me_2S, H_2/Lindlar's catalyst
acids	Ag_2O, $KMnO_4$, peracids
esters	$MeOH/H^+$

chromatography. Attempts to use this procedure to analyse mixtures of monoenes (such as are produced by partial catalytic hydrogenation) have met with only limited success. Oxidative cleavage does not distinguish between *cis*- and *trans*-isomers unless these are first separated from one another.

1.15 Fatty acid biosynthesis

A full account of the biosynthesis of organic molecules provides information on (i) the chemical pathway by which the natural product is derived from readily available starting material; (ii) the nature of the enzymes involved at each stage of the biosynthetic process; (iii) the regulatory procedures; and (iv) the site of these reactions in the plant, animal, or micro-organism. This report will be limited to the first of these.

The major biosynthetic pathways for fatty acids are summarized in Figure 1.3, and involve up to five processes. These are: (i) the *de novo* synthesis of saturated acids from acetate, itself a product of carbohydrate and fat catabolism; (ii) chain elongation; (iii) Δ9-desaturation to produce monoenes; (iv) further desaturation occurring in plant systems; and (v) further desaturation occurring in animal systems.

1.15.1 de novo *Synthesis of saturated acids*

Apart from a few minor compounds discussed later, all the carbon atoms present in fatty acids come from the two carbon atoms in acetate with one half derived from the original methyl carbon atom and the other half from the carboxyl carbon atom. This involvement of a C_2 unit explains the predominance of fatty acids with an even number of carbon atoms. However, the main extending unit is not acetate but the more reactive malonate which is produced from acetate and carbon dioxide in the presence of acetyl-CoA carboxylase thus:

$$CH_3COSCA + CO_2 \rightarrow HO_2CCH_2COSCoA$$

In the subsequent condensation of acetate (or other fatty acid) with

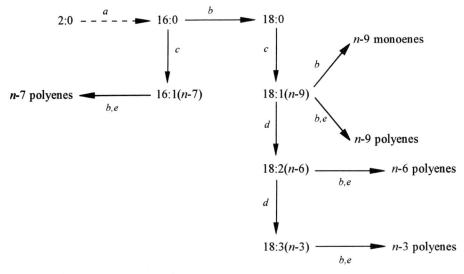

Figure 1.3 Major biosynthetic pathways for fatty acids. *a*, *de novo* synthesis of saturated acids from acetate. *b*, chain elongation. *c*, $\Delta 9$-desaturation. *d*, further desaturation in plant systems. *e*, further desaturation in animal systems.

malonate carbon dioxide is lost and the carbon atom in this molecule, which does not appear in the final product, is the same carbon atom as was added from carbon dioxide. Thus, despite the involvement of malonate, all the carbon atoms in the fatty acids come from acetate. Acetate and the malonate derived from it are sometimes replaced by propionate and the methylmalonate derived from this C_3 acid. These furnish branched methyl acids.

$$\begin{array}{cc} CH_3 & CH_3 \\ | & | \\ CH_2COSCoA & HOOCCHCOSCoA \\ \text{propionyl-CoA} & \text{methylmalonyl-CoA} \end{array}$$

The so-called acetate–malonate pathway is of primary importance and leads to three categories of natural products depending on which synthetic route is followed. Fatty acids are produced by the reductive pathway which will be described here, but acetate and malonate can also furnish both a wide range of phenolic compounds by cyclization of polyacetate and the important isoprenoids (terpenes and steroids) *via* mevalonic acid. This illustrates the observation that nature is economical both in the range of substrates employed and the reactions employed in natural product biosynthesis.

In the *de novo* pathway acetate and malonate react via a condensation and reduction cycle to produce first the C_4 acid (butanoic). Each repetition of the cycle adds two more carbon atoms until the fatty acid is finally detached from the enzyme system by a hydrolase. This is most often done at the C_{16} level, hence the importance of palmitic acid, but in the lauric oils the system appears to be optimized for lauric acid (12:0).

The acyl CoA esters of acetate and malonate are first transferred to acyl carrier protein (ACP) which is a thiol enzyme. These are then condensed, with loss of carbon dioxide, to give 3-oxobutanoate (a C_4 compound) which is reduced to a hydroxy compound (the 3R isomer), dehydrated to a *trans* 2-enoate, and reduced to butanoate. Through this sequence of reactions the starter acetate molecule has been extended by two carbon atoms to butanoate. The C_4 acid enters into the cycle again and reacts with malonate to give successively C_6, C_8, C_{10}, etc. acids until palmitate is formed. The complex of enzymes required for this sequence is described as fatty acid synthetase (FAS) (Figure 1.4).

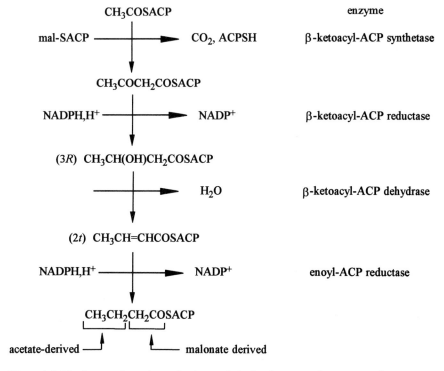

Figure 1.4 The first condensation-reduction cycle in the *de novo* pathway converting acetate to butanoate. (ACPSH = acyl carrier protein with a thiol group, $NADP^+$ and NADPH = nicotinamide adenine dinucleotide phosphate and its reduced form.)

When the usual C_2 starter (acetate) is replaced by other starters, still reacting with malonate, different products are obtained. For example:

CH₃COOH   →   even-chain acids, especially palmitate

CH₃CH₂COOH   →   odd-chain acids, especially heptadecanoate (C_{17})

(CH₃)₂CHCOOH*   →   even-chain iso acids, especially C_{18}

CH₃CH₂CH(CH₃)COOH*   →   odd-chain anteiso acids, especially C_{17}

⬠— COOH

                                 →   cyclopentene acids

*These are protein metabolites produced from valine and isoleucine, respectively.

Another small modification of the *de novo* pathway leads to the *n*-7 16:1 and 18:1 acids. This sequence does not require oxygen and is used by anaerobic micro-organisms. In the normal *de novo* sequence at the C_{10} level, 3-hydroxydecanoic acid is dehydrated to the Δ2*t* acid, then reduced and chain-extended further. In the modification dehydration gives the Δ3*c* acid and in the subsequent chain extension this double bond is not reduced but remains in the carbon chain so that 9-hexadecenoic acid is finally produced. Chain extension (see below) converts this to *cis*-vaccenic acid (11*c*-18:1).

→ 3-OH 10:0 → 10:1 (2*t*) → 10:0 – – – – – – – – – – – – – → 16:0
     ↓
  10:1 (3*c*) → 12:1 (5*c*) → 14:1 (7*c*) → 16:1 (9*c*) → 18:1 (11*c*)

1.15.2   Chain elongation

Chain elongation has many resemblances to the *de novo* synthetic route but differs in several important respects. The substrate – as ACP or CoA ester – is usually a preformed fatty acid which may be saturated or unsaturated. This reacts with acetyl or malonyl-CoA and after the usual condensation, reduction, dehydration, and a second reduction furnishes an acid with two additional carbon atoms which have been added at the carboxyl end of the molecule. This is the pathway by which palmitic acid is converted to stearic acid and to even longer-chain acids (up to C_{30}) and by which oleic acid is converted to long-chain monoenes. Although the double bond changes its position with respect to the carboxyl group there is no change with respect to the methyl group and all the acids produced from oleic acid will be *n*-9 monoenes (Figure 1.5). To develop a low-erucic rapeseed oil lines were developed in which the elongase normally present was much less active. The structural resemblance between ricinoleic (12-OH 18:1 9*c*) and lesquerolic acid (14-OH 20:1 11*c*) also suggests that the latter is produced from the former by chain elongation.

18:1 (9) $\longrightarrow$ 20:1 (11) $\longrightarrow$ 22:1 (13) $\longrightarrow$ 24:1 (15)

oleic   cetoleic   erucic   nervonic

$\longrightarrow$ 26:1 (17) $\longrightarrow$ 28:1 (19) $\longrightarrow$ 30:1 (21)

Figure 1.5 The *n*-9 family of monoene acids based on oleic acid (figure in parentheses indicates position of *cis* double bond with respect to the carboxyl group).

1.15.3   Desaturation to monoene acids

Unsaturated acids are produced by anaerobic and aerobic pathways of which the latter is the more usual. Desaturation in an anaerobic environment occurs by the modification of the *de novo* pathway already discussed (section 1.15.1).

The first double bond to be inserted in a saturated acyl chain is usually placed in the Δ9 position through stereospecific and regiospecific removal of the pro-*R* hydrogen atoms from C9 and from C10 to produce the *cis* (*Z*) alkene. With stearic acid, reaction occurs with a CoA ester, an ACP ester, or a phospholipid. Other Δ9 acids such as 9*c*-16:1 and 9*c*-14:1 are produced similarly from the appropriate saturated acids. The system operates in plants and animals, is oxygen dependent, and requires NADH or NADPH (the reduced form of nicotinamide adenine dinucleotide and its phosphate).

There must also be plant desaturases which can produce Δ5 and Δ3 acids (section 1.7). It is also necessary to explain the bioproduction of petroselinic acid (18:1 6*c*). Recent studies suggest that this is produced through a Δ4 desaturase and an elongase which together convert palmitic acid to petroselinic acid:

$$16:0 \xrightarrow{\text{4-desaturase}} 4c\text{-}16:1 \xrightarrow{\text{elongase}} 6c\text{-}18:1 \xleftarrow{\text{6-desaturase}}\!\!\!\!\times 18:0$$

1.15.4   Desaturation to polyene acids

The introduction of further double bonds after the first occurs by different pathways in the plant and animal kingdoms. In plants further double bonds are most commonly introduced between the existing double bond and the methyl group. These double bonds have *cis* (*Z*) configuration and take up a methylene-interrupted pattern. The most important is the C_{18} group oleate–linoleate–linolenate. Each of these is the precursor of a family of polyene acids. Linoleate and α-linolenate are essential fatty acids (EFA) since they and/or their metabolites are required for good health. As they cannot be made by animals they must be obtained from dietary vegetable sources.

$$\text{oleate} \xrightarrow{\text{desaturase}} \text{linoleate} \xrightarrow{\text{desaturase}} \text{α-linolenate}$$
$$\text{9}c\text{-18:1 }(n\text{-9}) \qquad \text{9}c\text{12}c\text{-18:2 }(n\text{-6}) \qquad \text{9}c\text{12}c\text{15}c\text{-18:3 }(n\text{-3})$$

Although it is not common, plants can also introduce double bonds between existing double bonds and the carboxyl group. A good example of this is the presence of γ-linolenic acid (6c9c12c-18:3) in evening primrose, borage, and blackcurrant seed oils (sections 1.5 and 3.2.1).

Animals, on the other hand, are apparently unable to introduce double bonds on the methyl side of the n-9 double bond and must derive their required linoleate and α-linolenate from plant-based dietary sources. But once available these acids can be desaturated and chain elongated (metabolized) to give other members of the n-6 and n-3 families (Figure 1.6). The existence of the 4-desaturase to convert DPA to DHA has been questioned and evidence presented to show that the final double bond is introduced by a 6-desaturase operating on a C_{24} acid thus:

$$22:5 \xrightarrow{\text{elongase}} 24:5 \xrightarrow{\text{6-desaturation}} 24:6 \xrightarrow{\beta\text{-oxidation}} 22:6$$
$$(7,10,13,16,19) \quad (9,12,15,18,21) \quad (6,9,12,15,18,21) \quad (4,7,10,13,16,19)$$

EFA deficiency can be observed in experimental animals fed on a diet deficient in linoleic and α-linolenic acids. It is seldom observed in human subjects except perhaps in certain disease conditions. It is apparent from the condition of the skin but can be measured biochemically from the ratio of Mead's acid (20:3 n-9) derived from oleic acid to arachidonic acid (20:4 n-6) derived from linoleic acid. This has been described as the triene–tetraene ratio and used as a biochemical index of EFA states. It has long been held that a ratio greater than 0.4 was diagnostic of EFA deficiency but recent work suggests that a lower figure should be used.

1.16   Fatty acids – chemical synthesis

1.16.1   Introduction

Most of the common fatty acids can be isolated from natural sources and can be purchased from chemical suppliers at purity levels of 90–99% and above. Some acids are available on a commercial scale at purities of about 70 to 95%. Sometimes, however, laboratory synthesis is necessary as in the following examples:

1. acids which are not readily isolated from natural sources because there is no easy access to a rich source: for example heptadecanoic acid (17:0) or myristoleic (9-tetradecenoic) acid;

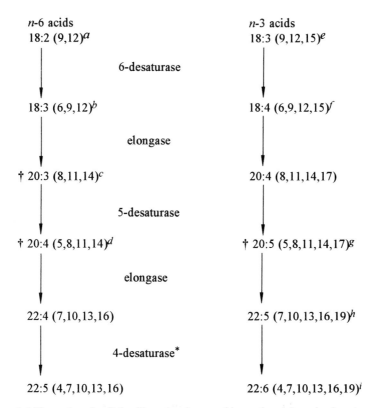

Figure 1.6 The *n*-6 and *n*-3 families of polyene acids produced in animal systems from linoleate and α-linolenate, respectively. *a*, linoleic. *b*, γ-linolenic. *c*, dihomom-γ-linolenic. *d*, arachidonic. *e*, α-linolenic. *f*, stearidonic. *g*, eicosapentaenoic (EPA). *h*, docosapentaenoic (DPA). *i*, docosahexaenoic (DHA).

*see text
†source of prostaglandins and other eicosanoid metabolites

2. acids which are not known to occur naturally such as many stereoisomers of octadecenoic acid;
3. acids required in an isotopically labelled form (^{2}H, ^{3}H, ^{13}C, ^{14}C): such acids are needed for the study of reaction mechanism or of biochemical processes.

Some acids can be made by chain extension of readily available starting acids. Standard procedures exist for extension by one, two, five, or six carbon atoms. These work well with saturated acids and some can be applied to monoene and to even more unsaturated acids. For example, heptadecanoic acid can be made from palmitic acid by C_1 extension or from lauric acid by C_5 extension. Both starting materials are easily obtained but the latter route has the advantage that it is easier to separate product (C_{17}) from the C_{12} starting material than from the C_{16} homologue.

$$RCO_2Me \xrightarrow{\text{i}} RCH_2OH \xrightarrow{\text{ii}} RCH_2OMs \xrightarrow{\text{iii}} RCH_2CH(CO_2Et)_2 \xrightarrow{\text{iv, v, vi}}$$

18:3 (*n*-6)   20:3 (*n*-6)

$RCH_2CH_2CO_2Me$

20:3 (*n*-6)

Figure 1.7 Conversion of methyl-γ-linolenate to its C_{20} homologue. (i) LiAlH$_4$; (ii) methanesulphonate (bromide could also be used), MsCl, NEt$_3$; (iii) NaCH(CO$_2$Et)$_2$; (iv) KOH, H$_2$SO$_4$; (v) heat at 165°C; (vi) MeOH,H$_2$SO$_4$.

$$Cl(CH_2)_5Cl \xrightarrow{\text{i}} I(CH_2)_5Cl \xrightarrow{\text{ii}} HC{\equiv}C(CH_2)_5Cl \xrightarrow{\text{iii}}$$

$$CH_3(CH_2)_9C{\equiv}C(CH_2)_5Cl \xrightarrow{\text{iv,v}} CH_3(CH_2)_9C{\equiv}C(CH_2)_5COOMe \xrightarrow{\text{vi}}$$

$$CH_3(CH_2)_9C{\equiv}C(CH_2)_5COOH \xrightarrow{\text{vii or viii}} CH_3(CH_2)_9CH{=}CH(CH_2)_5COOH$$

Figure 1.8 Synthesis of 7-octadecenoic acid. (i) NaI, COMe$_2$; (ii) NaC≡CH, NH$_3$; (iii) NaNH$_2$, NH$_3$, CH$_3$(CH$_2$)$_9$Br; (iv) KCN, DMSO; (iv) MeOH, H$^+$; (vi) KOH, H$_3$O$^+$; (vii) H$_2$, Pd to give the *cis* isomer; (viii) Li, NH$_3$ to give the *trans* isomer.

Dihomo-γ-linolenic ester (20:3 *n*-6) is easily obtained by a C_2 chain elongation (with ethyl malonate) from γ-linolenic ester, itself isolated from a natural source (Figure 1.7).

The preparation of other acids by industrial processes are covered in sections 7.5.3 and 7.11.

1.16.2 Synthesis via acetylenic intermediates

For unsaturated acids the two most common reaction sequences involve the use of acetylenic intermediates or the Wittig reaction.

The first of these depends on the fact that acetylene (ethyne) can be alkylated once or twice and that the triple bond can be partially reduced stereospecifically to give *cis*- or *trans*-olefinic compounds. A simple example to produce a mono-olefinic acid is given in Figure 1.8.

This reactivity of alkynes can be successfully extended to give polyenes in a number of ways. One important route to methylene-interrupted polyenes involves reaction between appropriate propargyl bromides and ethynyl compounds as organometallic derivatives – mainly sodium, lithium, or magnesium – in the presence of a cuprous salt:

$$RC{\equiv}CCH_2Br + BrMgC{\equiv}CR' \xrightarrow{Cu^+} RC{\equiv}CCH_2C{\equiv}CR'$$

Further unsaturation may already be present in one or both of groups R and R'. Poly-ynoic acids are crystalline solids which must be purified by repeated crystallization before partial reduction in the presence of Lindlar's catalyst. The resulting polyene acids are purified, if required, by silver ion chromatography to remove undesired over-reduced and/or *trans*-olefinic isomers. The example detailed in Figure 1.9 leads to a C_{19} tetraene bromide which can be used as a precursor for labelled arachidonic acid.

1.16.3   Synthesis by the Wittig reaction

In the Wittig reaction an alkyl halide, as its phosphonium salt, reacts with a base to produce the ylid which is then condensed with an aldehyde:

$$RCH_2Br \xrightarrow{Ph_3P} RCH_2\overset{+}{P}PH_3\overset{-}{Br} \xrightarrow{base} \underset{ylid}{RCH=PPh_3} \xrightarrow{R'CHO} RCH$$

The product may be a mixture of *cis*- and *trans*-isomers but each of these can be optimized by appropriate selection of reaction conditions. The *cis*-isomers are dominant when the reaction is carried out at low temperature ($-100°C$), at high dilution, and in the absence of Li^+. Sodium bistrimethyl-silylamide [$NaN(SiMe_3)_2$] is recommended as the base. If necessary the final product can be purified by silver ion chromatography. This is illustrated (Figure 1.10) with a reaction sequence producing an isomer of arachidonic acid. This has been used to prepare the 8,8,9,9-d_4 isomer.

$$HC\equiv CCH_2CHC(OEt)_2 \xrightarrow{i,ii} RC\equiv CCH_2C\equiv CCH_2CH(OEt)_2 \xrightarrow{iii}$$

$$RCH=CHCH_2CH=CHCH_2CH(OEt)_2 \xrightarrow{iv,v} RCH=CHCH_2CH=CHCH_2CH=CBr_2 \xrightarrow{vi}$$

$$RCH=CHCH_2CH=CHCH_2C\equiv CH \xrightarrow{vii} R(CH=CHCH_2)_2(C\equiv CCH_2)_2(CH_2)_2Othp \xrightarrow{iii}$$

$$R(CH=CHCH_2)_4(CH_2)_2Othp \xrightarrow{iv,v} R(CH=CHCH_2)_4CH_2CH_2Br$$

Figure 1.9 Synthesis of C_{19} bromide as a precursor for arachidonic acid. (i) EtMgBr, THF; (ii) RC≡CCH$_2$Br, CuBr, R=CH$_3$(CH$_2$)$_4$; (iii) P-2 Ni, H$_2$; (iv) aq CF$_3$COOH; (v) PPh$_3$, CBr$_4$, CH$_2$Cl$_2$; (iv) BuLi, THF; (vii) BrCH$_2$C≡C(CH$_2$)$_3$Othp, CuBr, THF (THF = tetrahydro-furan, thp = tetrahydropyranyl).

$$CH_3CH_2(CH=CHCH_2)_3CD_2CD_2CH_2CHO + \overset{-}{Br}\overset{+}{Ph_3}\overset{}{P}(CH_2)_4COOH \xrightarrow{i,ii}$$

$$CH_3CH_2(CH=CHCH_2)_3CD_2CD_2CH_2CH=CH(CH_2)_3COOCH_3$$

Figure 1.10 Synthesis of 20:4 (5,11,14,17)-8,8,9,9d$_4$. (i) NaN(SiMe$_3$)$_2$, THF; (ii) CH$_3$OH, H$^+$ (THF = tetrahydrofuran).

$$CH_3(CH_2)_4CHO \xrightarrow{\text{i,ii}} CH_3(CH_2)_4CH{=}CHCH_2CHO \xrightarrow{\text{i,ii}}$$

$$CH_3(CH_2)_4(CH{=}CHCH_2)_2CHO \xrightarrow{\text{i,ii}} CH_3(CH_2)_4(CH{=}CHCH_2)_3CHO \xrightarrow{\text{iii}}$$

$$CH_3(CH_2)_4(CH{=}CHCH_2)_4(CH_2)_2COOH$$

Figure 1.11 Synthesis of arachidonic acid. (i) $\bar{B}rPh_3\overset{+}{P}CH_2CH_2CH(OPri)_2$, NaN(SiMe$_3$)$_2$, THF, HMPA; (ii) H$_2$O, TsOH; (iii) $\bar{B}rPh_3\overset{+}{P}(CH_2)_4COOH$, NaN(SiMe$_3$)$_2$, THF, HMPA (THF = tetrahydrofuran, HMPA = hexamethylphosphoramide).

The ylid produced from $\bar{B}rPh_3\overset{+}{P}CH_2CH_2CH(OPr^i)_2$ – itself obtained from propenal (acrylic aldehyde) – is a convenient synthon which is used three times to produce arachidonic acid from hexanal in four steps with an overall yield of 58% (Figure 1.11).

1.16.4 Isotopically labelled acids

Acids with isotopic hydrogen (^{2}H, ^{3}H) or carbon (^{13}C, ^{14}C) are produced by modification of the methods which have already been described or by incorporation into small molecules which are then involved in synthesis. Such acids are required for the study of reaction mechanisms or lipid biosynthesis and metabolism. The compounds and products derived from them can be studied on the basis of their radioactivity (^{3}H, ^{14}C) by mass spectrometry or by NMR spectrometry, as is most appropriate. Deuterium-labelled compounds have sometimes been prepared in quantities sufficient for nutritional studies since these materials can be safely ingested.

The deuterium (^{2}H) label is usually incorporated into an appropriate substrate by exchange or reduction (hydrogenation). Hydrogen on carbon adjacent to an aldehyde or ketone function is replaced by ^{2}H from ^{2}H$_2$O in the presence of a mildly acidic or basic catalyst which promotes enolization.

$$RCH_2CHO \xrightarrow[\text{pyridine}]{^2H_2O} RC^2H_2CHO$$

The similar reaction with carboxylic acids requires more vigorous conditions.

$$RCH_2COOH \xrightarrow[\text{150–200}]{^2HO^-} RC^2H_2COOH$$

Useful reduction processes include partial catalytic reduction of alkynes with deuterium, reduction of alkenes with tetradeuteriohydrazine, and reduction of carbonyl compounds with lithium aluminium deuteride:

$$RC\equiv CR' \xrightarrow[\text{Lindlar's catalyst}]{^{2}H_2} RC^{2}H=C^{2}HR'$$

$$RCH=CHR' \xrightarrow{N_2{}^{2}H_4} RC^{2}HHC^{2}HHR'$$

$$RCOOMe \xrightarrow{LiAl^{2}H_4} RC^{2}H_2OH$$

Catalytic hydrogenation with a heterogeneous catalyst is not a satisfactory procedure because reduction is accompanied by extensive exchange between hydrogen and deuterium leading to a mixture of polydeuterated molecules.

These same methods can be applied to the introduction of tritium. The reduction of mesylates (or tosylates) has also proved useful:

$$>CHOH \xrightarrow{MsCl, NEt_3} >CHOMs \xrightarrow{LiAl^{3}H_4} >CH^{3}H$$
$$\text{methanesulphonate}$$

When preparing acids labelled with ^{13}C or ^{14}C the isotope should be introduced as late in the synthetic sequence as possible because of its high cost. The most common carriers for the isotopic carbon are potassium cyanide and metallic carbonates (as a source of labelled carbon dioxide). Simplified examples are set out:

$$RCl \xrightarrow{K^{14}CN, DMSO} R^{14}CN \xrightarrow{MeOH, H^+} R^{14}COOMe$$

$$RBr \xrightarrow[{^{14}CO_2, H_3O^+}]{Mg, Et_2O,} R^{14}COOH \xrightarrow{MeOH, H^+} R^{14}COOMe$$

$$RCH_2OH \xrightarrow[NEt_3]{MsCl} RCH_2OMs \xrightarrow[MeOH,H^+]{K^{14}CN} RCH_2{}^{14}COOMe$$

Bibliography

Badami, R.C. and K.B. Patil (1982) Structure and occurrence of unusual fatty acids in minor seed oils, *Prog. Lipid Res.*, **19**, 119–153.

Emken, E.A. and H.J. Dutton (eds) (1979) *Geometrical and Positional Fatty Acid Isomers*, AOCS, Champaign, USA (metabolism).

Gunstone, F.D. (1980) Natural oxygenated acids, in *Fats and Oils: Chemistry and Technology* (eds R.J. Hamilton and A. Bhati), Applied Science, London, pp. 47–58.

Gunstone, F.D. (1990) Fatty acids – structural identification, in *Methods of Plant Biochemistry* (eds J.L. Harwood and J.R. Bowyer), Academic Press, London, pp. 1–17.

Gunstone, F.D. *et al.* (eds) (1994) *The Lipid Handbook*, 2nd edn, Chapman and Hall, London.

Gurr, M.I. and J.L. Harwood (1991) *Lipid Biochemistry – an Introduction*, 4th edn, Chapman and Hall, London (fatty acid structure and metabolism).

Harwood, J.L. (1981) The synthesis of acyl lipids in plant tissues, *Prog. Lipid Res.*, **18**, 55–86.

Harwood, J.L. and N.J. Russell (1984) *Lipids in Plants and Microbes*, George Allen and Unwin, London (biosynthesis of fatty acids).

Heitz, M.P., A. Wagner, and C. Mioskowski (1989) *J. Org. Chem.*, **54**, 500–503.

Hopkins, C.Y. (1973) Fatty acids with conjugated unsaturation, in *Topics in Lipid Chemistry* (ed. F.D. Gunstone), Vol. 3, pp. 37–87.

Lie Ken Jie, M.S.F. (1979) Synthesis of three-, five-, and six-membered carbocyclic and furanoid fatty acids, *Chem. Phys. Lipids*, **24**, 407–430.

Lie Ken Jie, M.S.F. (1993) The synthesis of rare and unusual fatty acids, *Prog. Lipid Res.*, **32**, 151–194.

Miwa, T.K., K.L. Mikolajczak, F.R. Earle and I.A. Wolff (1960) *Anal. Chem.*, **32**, 1739–1742.

Polgar, N. (1971) Natural alkyl-branched long-chain acids, in *Topics in Lipid Chemistry* (ed. F.D. Gunstone), Vol. 2, pp. 207–246.

Pryde, E.H. (1979) Natural fatty acids and their sources, in *Fatty Acids* (ed. E.H. Pryde), AOCS, Champaign, USA, pp. 1–28.

Rakoff, H. (1982) Preparation of fatty acids and esters containing deuterium, *Prog. Lipid Res.*, **21**, 225–254.

Rezanka, T. (1989) Very-long-chain fatty acids from the animal and plant kingdoms, *Prog. Lipid Res.*, **28**, 147–187.

Smith, C.R. (1971) Occurrence of unusual fatty acids in plants, *Prog. Chem. Fats and Other Lipids*, **11**, 137–177.

Smith, C.R. (1979) Unusual seed oils and their fatty acids, in *Fatty Acids* (ed. E.H. Pryde), AOCS, Champaign, USA, pp. 29–47.

Sommerfield, M. (1983) *Trans* unsaturated fatty acids in natural products and processed foods, *Prog. Lipid Res.*, **22**, 221–233.

Stumpf, P.K. (1989) Biosynthesis of fatty acids in higher plants, in *Oil Crops of the World* (eds G. Robbelen, R.K. Downey, and A. Ashri) McGraw-Hill, New York, pp. 38–62.

Tulloch, A.P. (1983) Synthesis, analysis, and application of specifically deuterated lipids, *Prog. Lipid Res.*, **22**, 235–256.

2 Lipids – Nomenclature, structure, biosynthesis, and chemical synthesis

2.1 Introduction

As already stated the author accepts and uses the following definition of lipids: "Lipids are compounds based on fatty acids or closely related compounds such as the corresponding alcohols or the sphingosine bases".

The fatty acids discussed in the previous chapter seldom occur in their free state but usually as esters and sometimes as amides (Table 2.1).

Esters contain acid and alcohol components and the most common alcohol present in lipids is glycerol (propane-1,2,3-triol). Important glycerolipids include the acylglycerols, the phospholipids, and the glycosylglycerides. Other alcohols which occur less commonly in lipids in an acylated form include carbohydrates (sugar esters), sterols (sterol esters), and long-chain alcohols (wax esters).

The amides are derivatives of long-chain acids with amines such as the sphingosine bases or some amino acids.

Before detailing these it is useful to explain some terms which occur frequently. Simple or neutral lipids are triacylglycerols, some ether lipids, sterol esters, and wax esters. Complex or polar lipids include phosphoglycerides, glycosyldiacylglycerols, and sphingolipids. Phospholipids are derivatives of phosphoric (or less commonly phosphonic) acid, glycolipids contain one or more carbohydrate units, and sulpholipids contain sulphur.

Glycerol is a prochiral molecule with the prochiral carbon atom carrying two CH_2OH groups. When these differ, as when they are acylated with different fatty acids, then the molecule is chiral and exists in two enantiomeric forms.

To designate the stereochemistry of glycerol-containing components, the carbon atoms are stereospecifically numbered (*sn*). When the glycerol molecule is represented by a Fischer projection with the secondary hydroxyl to the left of the central (prochiral) carbon atom, then the carbon atoms are numbered 1, 2, and 3 from top to bottom. Molecules which are stereospecifically numbered in this fashion have the prefix *sn*- immediately preceding the term 'glycerol'. The prefix *rac*- in front of the full name shows that the compound is an equal mixture of both enantiomers. When *x*- is used the configuration is unknown or unspecified.

$$
\begin{array}{ccc}
\text{CH}_2\text{OH} & sn\text{-}1 & \alpha \\
\text{HO} \underline{\quad\quad} \text{H} & sn\text{-}2 & \beta \\
\text{CH}_2\text{OH} & sn\text{-}3 & \alpha \text{ or } \alpha'
\end{array}
$$

Table 2.1 Types of lipids

esters of:	
glycerol	acylglycerols (glycerides)
	phosphoglycerides (PA, PC, PE, PS, PI, PG)[a]
	glycosylglycerides (MGDG, DGDG)[a]
other alcohols	
diols	–
sugars	sugar esters
long-chain alcohols[b]	wax esters
sterols	sterol esters
amides of:	
long-chain bases (spingosine etc.)	ceramides, cerebrosides, gangliosides
taurine	–
serine	–

[a]PA phosphatidic acids, PC phosphatidylcholines, PE phosphatidylethanolamines, PS phosphatidylserines, PI phosphatidylinositols, PG phosphatidylglycerols, MGDG monogalactosyldiacylglycerols, DGDG digalactosyldiacylglycerols.
[b]Also form ether lipids (see section 2.5).

Any glycerol lipid will be chiral when the substituents at the *sn*-1 and *sn*-3 positions are different. Mirror image molecules or enantiomers possess opposite but equal rotations. However, if both substituents are long-chain acyl groups then the optical rotations are very small and may be difficult to observe and measure.

2.2 Acylglycerols

The major reserve lipids are triacylglycerols (formerly known as triglycerides). Monoacylglycerols and diacylglycerols may also be present as minor components. However, these are important intermediates in the biosynthesis and catabolism of triacylglycerols and other classes of lipids. They are also prepared on a large scale for use as surface-active agents (see sections 8.3.3 and 10.5).

2.2.1 *Monoacylglycerols* (monoglycerides)

These are fatty acid monoesters of glycerol and exist in two isomeric forms thus:

$$CH_2OCOR \qquad CH_2OH \qquad CH_2OH$$

$$HO \overline{\qquad} H \qquad HO \overline{\qquad} H \qquad RCOO \overline{\qquad} H$$

$$CH_2OH \qquad CH_2OCOR \qquad CH_2OH$$

1- and 3- monoacyl-*sn*-glycerols   2-monoacyl-*sn*-glycerol

α-monoglyceride   β-monoglyceride

Pure isomers readily change to a 90:10 mixture of the α- and β-isomers. This rearrangement is promoted by acid or alkali and is an example of transesterification (section 8.3.4).

2.2.2 *Diacylglycerols* (diglycerides)

These are fatty acid diesters of glycerol and they also exist in two isomeric forms:

$$CH_2OCOR \qquad CH_2OH \qquad CH_2OCOR$$

$$R'COO \overline{\qquad} H \qquad R'COO \overline{\qquad} H \qquad HO \overline{\qquad} H$$

$$CH_2OH \qquad CH_2OCOR \qquad CH_2OCOR'$$

1,2- and 2,3-diacyl-*sn*-glycerols   1,3-diacyl-*sn*-glycerol

α β-diglycerides   αα'-diglycerides

They readily form an equilibrium mixture, with the 1,3-diacylglycerol being the more stable.

2.2.3 *Triacylglycerols* (triglycerides)

These are fully acylated derivatives of glycerol:

$$CH_2OCOR^1$$

$$R^2COO \overline{\qquad} H$$

$$CH_2OCOR^3$$

Table 2.2 Triacylglycerols containing only palmitic (P) and oleic (O) acids

PPP	POP $\begin{bmatrix} OPP \\ PPO \end{bmatrix}$[a]	OPO $\begin{bmatrix} POO \\ OOP \end{bmatrix}$[a]	OOO

[a]Unsymmetrical triacylglycerols in enantiomeric pairs.

Table 2.3 The relation between the number of acids and the maximum number of triacylglycerols which can be formed from them

	Number of acids			
triacylglycerols	5	10	20	n
all isomers distinguished	125	1000	8000	n^3
excluding optical isomers	75	400	4200	$(n^3 + n^2) \div 2$
no isomers distinguished	35	220	1540	$(n^3 + 3n^2 + 2n) \div 6$

It is unusual for a natural triacylglycerol to have only one acid although this does happen, e.g. triolein in olive oil and tripalmitin in palm oil. More usually, two or three different acids are present. The number of possible triacylglycerols rises very quickly with the number of acids present in the fatty acid pool. With only two acids there are eight triacylglycerols (including enantiomers, Table 2.2). With three acids this number is 27 and it rises very rapidly thereafter (Table 2.3). The triacylglycerol composition of some natural oils is discussed in sections 3.5 and 5.2.4.

On hydrolysis with aqueous acid or alkali the triacylglycerols are split into glycerol and a mixture of fatty acids. Enzymic hydrolysis usually occurs with some degree of regiospecificity. Most commonly this involves deacylation of the α-positions (*sn*-1 and *sn*-3) with no reaction occurring at the β-position so that the products are fatty acids from the α positions and 2-monoacylglycerols:

[1, 2, and 3 represent the mixed fatty acids present at each of these positions]

This selective deacylation occurring with a lipase (designated lipolysis) is an important step in the metabolism of triacylglycerols and is also the basis of an analytical procedure for examining fatty acid distribution (sections 5.2.6 and 9.2).

Some glycerol esters have more than three acyl groups. This is possible when the lipid contains hydroxy acids in which the hydroxyl group can also be acylated. In some cases, such as castor oil with about 90% of ricinoleic acid (12-hydroxyoleic), there is no evidence of any species other than triacylglycerols but other examples are known with up to six acyl groups. One example is provided by the glycerol esters in *Sapium sebiferum* seed oil (stillingia oil). One of its major glycerides has the structure shown below in which a C_8 hydroxy allenic acid is attached to a C_{10} acid:

H_2C-len
|
lin-CH
|
$CH_2OCO(CH_2)_3CH=C=CHCH_2OCO(CH=CH)_2(CH_2)_4CH_3$

lin = linoleic (18:2)   len = linolenic (18:3)

2.3 Glycosyldiacylglycerols

This class of glycolipids is present in plants (especially in the chloroplast) and in bacteria. Diacylglycerols are linked to one or more sugar units (especially galactose) at the *sn*-3 position. They are usually rich in polyunsaturated C_{18} acids. Typical examples are the monogalactosyldiacylglycerols (MGDG) and the digalactosyldiacylglycerols (DGDG).

	sugar unit	
1		
2	galactose	MGDG
O sugar	galactose-galactose	DGDG

2.4 Phosphoglycerides

There are several types of phosphoglycerides (Table 2.4). These are based on the phosphatidic acids which are diacyl derivatives of 3-glycerophosphoric acid. The phosphatidic acids are monoesters of the tribasic phosphoric acid. The remaining phospholipids are diesters, i.e. a second hydroxy compound is associated with the molecule (Table 2.4).

$$CH_2OH$$
$$|$$
$$HOCH \quad O$$
$$| \quad ||$$
$$CH_2OPOH$$
$$|$$
$$OH$$

3-glycerophosphoric acid

$$CH_2OCOR^1$$
$$|$$
$$R^2COOCH \quad O$$
$$| \quad ||$$
$$CH_2OPOX$$
$$|$$
$$OH$$

phosphatidic acids (PA, X = H)

Table 2.4 The major glycerophospholipids

Name	Abbreviation	X^a
phosphatidylcholines	PC	$CH_2CH_2\overset{+}{N}Me_3$
phosphatidylethanolamines[b]	PE	$CH_2CH_2\overset{+}{N}H_3$
phosphatidylserines	PS	$CH_2CH(\overset{+}{N}H_3)COO^-$
phosphatidylinositols	PI	$C_6H_{11}O_5$
phosphatidylglycerols	PG	$CH_2CH(OH)CH_2OH$

[a]For X see the structure of the phosphatidic acids in the text.
[b]Also occur as esters of phosphonic acid (H_3PO_3).

The composition of some natural phospholipids is discussed in sections 2.4 and 3.8. Many phospholipids contain saturated acids at the *sn*-1 position and an unsaturated fatty acid at the *sn*-2 position but this is not an invariable rule. The phosphoglycerides, with two lipophilic chains and a hydrophobic head group, are important structural lipids. They are essential parts of lipid bilayers which make up cell membranes.

Each phosphoglyceride has four ester groups and their hydrolysis is illustrated for phosphatidylcholines. Complete hydrolysis occurs with aqueous acid and the products are glycerol, fatty acids, phosphoric acid, and choline. With alkali the acyl groups are easily hydrolysed but the phosphate ester bonds react more slowly so that hydrolysis furnishes fatty acids and glycerophosphorylcholines (GPC) which are then slowly hydrolysed to choline and glycerophosphoric acid. This last is a mixture of α- and β-isomers because of the migration of phosphoric acid among the hydroxyl groups. Before this reaction was fully understood it led to the idea that the phosphoglycerides were mixtures of α- and β-isomers.

Enzymic hydrolysis is more selective and gives rise to a range of interesting products. Phospholipase A1 causes deacylation only at the *sn*-1 position liberating the fatty acids from this position and leaving a lysophosphatidylcholine. Phospholipase A2 behaves similarly at the *sn*-2 position. Phospholipase B is now recognized as a mixture of the A1 and A2 enzymes and removes fatty acids from both positions leaving glycerophosphorylcholine. Phospholipase C reacts at one of the phosphate ester bonds to give 1,2-diacylglycerols and phosphorylcholine, while phospholipase D cleaves the other phosphate ester bond to give choline and phosphatidic acids (Table 2.5).

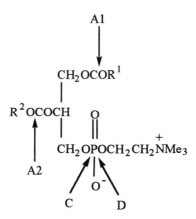

These stereospecific processes have been exploited in a number of synthetic procedures. Starting with natural phosphatidylcholines, for example, it is possible to remove the fatty acids from the *sn*-1 and/or *sn*-2 positions and to replace these with an acid of choice, thereby retaining the stereochemistry of the natural product. Also, the phosphatidylcholines can be converted to phosphatidic acids by reaction with phospholipase D and then esterified as required to produce other phosphoglycerides (section 2.9.8).

Table 2.5 Enzymic hydrolysis of phosphatidylcholines[a]

Phospholipase	Lipolysis products	Source
A1	fatty acids (*sn*-1), lyso PC	snake venom
A2	fatty acids (*sn*-2), lyso PC	snake venom
B	fatty acids (*sn*-1 and 2), GPC	–
C	DAG, phosphorylcholine	bacteria (e.g. *Clostridia*)
D	PA, choline	most plant tissues

PC phosphatidylcholines, GPC glycerophosphorylcholines, DAG diacylglycerols, PA phosphatidic acids.
[a]These enzymes are also effective with phosphatidylethanolamines.

2.5   Ether lipids

Four types of monoether lipids have been recognized depending on whether they are related to triacylglycerols or phosphoglycerides and on whether the ether group contains or does not contain a Δ1 double bond (*trans*). The ether group is generally in the *sn*-1 position though some diethers (*sn*-1 and *sn*-2) are also known.

Compounds of type 1 occur in some marine oils. When they are hydrolysed they yield fatty acids (2 moles) and glycerol ethers. The alkyl chain is commonly 16:0, 18:0, or 18:1 and the glycerol ethers (which are diols) derived from these are known as chimyl alcohol (16:0), batyl alcohol (18:0), and selachyl alcohol (9c-18:1). The vinyl ether group present in compounds of types 2 and 4 resists reaction with alkali and with lithium aluminium hydride but is hydrolysed under acidic conditions to give an aldehyde.

$$\textbf{2} \;\rightarrow\; \textbf{glycerol} + \textbf{fatty acids} + [R^1CH=CHOH] \rightleftharpoons R^1CH_2CHO$$

Ethanolamine derivatives of type 4 occur in animal tissue and were originally called plasmalogens to reflect the production of aldehydes. Their biochemical significance is not understood.

Compounds designated platelet-activating factor (PAF) have biological activity (aggregation, inflammatory, edemic) at very low concentrations and are specifically bound to receptors in platelets. They are 1-*O*-alkyl-2-acetyl-*sn*-glycerol-3-phosphocholines:

The diether lipid, diphytanylglycerol, is an important component of *Archaebacteria*:

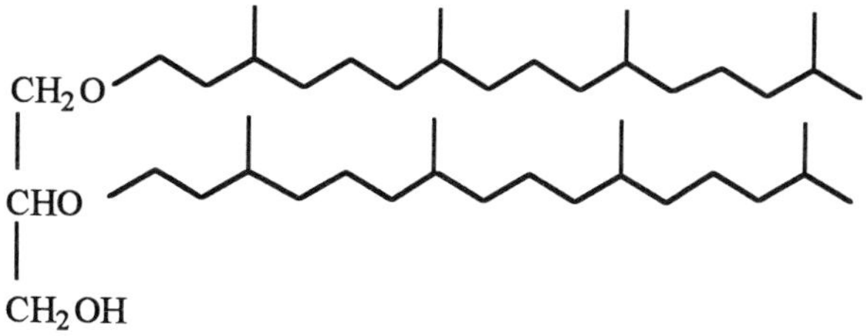

CH$_2$O

|

CHO

|

CH$_2$OH

2,3-di-O-phytanyl-*sn*-glycerol

2.6 Acyl derivatives of other alcohols

All the lipids discussed so far have been esters of glycerol. There remain to be detailed some esters of fatty acids with other alcohols.

Ester waxes (section 3.10) are derivatives of long-chain acids and long-chain alcohols such as occur in fish (sperm whale oil, orange roughy oil), animals (beeswax, wool wax), and plants (carnauba, jojoba). Canauba wax, for example, contains about 36% of ester wax in addition to other long-chain compounds. The esters are mainly C$_{46}$–C$_{54}$ molecules resulting from C$_{16}$–C$_{20}$ acids and C$_{30}$–C$_{34}$ alcohols. Jojoba oil contains both monoene acids [18:1 (6%), 20:1 (35%), and 22:1 (7%)] and monoene alcohols [20:1 (22%), 22:1 (21%), and 24:1 (4%)].

Esters (and ethers) based on ethanediol and propane-1,3-diol have also been identified. The latter are probably widely distribued at low levels but are difficult to detect in the presence of glycerol esters. Sterols, sugars, and other natural alcohols also occur as esters with long-chain alcohols. This acylation leads to a reduction in water solubility and an increase in fat solubility.

Cutins and suberins are plant polymers produced from mono-, di- and trihydroxy C$_{16}$ and C$_{18}$ acids.

2.7 Sphingolipids

The component units of sphingolipids are fatty acids, often including 2-hydroxyacids, bound as amides to long-chain amines. These amines also have two or three hydroxyl groups one of which is linked to one or more sugar units or to a phosphoester unit.

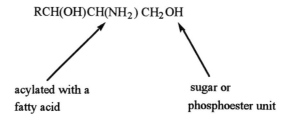

RCH(OH)CH(NH$_2$) CH$_2$OH

acylated with a   sugar or
fatty acid   phosphoester unit

The long-chain bases (sphingoids) are mainly C$_{18}$ or C$_{20}$ compounds and
the structure of two such bases is given:

$$t(E)\quad (R)\quad (S)$$

sphingosine $CH_3(CH_2)_{12}CH=CHCH(OH)CH(NH_2)CH_2OH$
(4t-sphingenine)

$$(R)\quad (S)\quad (S)$$

phytosphingosine $CH_3(CH_2)_{13}CH(OH)CH(OH)CH(NH_2)CH_2OH$

Their *N*-acyl derivatives are known generally as ceramides and typical
glycosphingolipids (gangliosides, cerebrosides) and a sphingomyelin are
set out in Table 2.6.

There are several diseases in which sphingolipids accumulate in various
organs and tissues through enzymic defects in sphingolipid metabolism.
These are known generally as sphingolipidoses and include diseases
associated with the names of Tay-Sachs, Fabry, Gaucher, Farber,
Niemann-Pick, and Krabbe.

Although structurally different, the phosphoglycerides and sphingolipids
resemble each other in that both contain two long chains and a polar head
group and are conveniently located in lipid bilayers.

2.8 Lipid biosynthesis

This simplified account of lipid biosynthesis is prefaced by some general
points:

Table 2.6 The structure and occurrence of some sphingolipids

Name	R^1	R^2	Occurrence
sphingenine	H	H	–
ceramides	COR'	H	skin
cerebrosides	COR'	glu or gal	brain, spleen, ganglion cells
gangliosides	COR'	glu-gal-sialic acid	myelin sheath of brain and nerve
sphingomyelin	COR'	PO$_3$CH$_2$CH$_2$ÑMe$_3$	animal membranes

RCH=CHCH(OH)CHCH$_2$OR^2
|
NHR^1

1. The glycerol present in glycerolipids is derived from glycerol, glyceralde-
   hyde (2,3-dihydroxypropanal), or dihydroxyacetone (1,3-dihydroxy-
   propan-2-one), all of which are products of carbohydrate metabolism.
2. Many lipids can be produced by more than one pathway.
3. Apart from the ether lipids, all glycerolipids are made from phosphatidic
   acids or 1,2-diacylglycerols which are themselves interconvertible
   (Figure 2.1).
4. Structural units to be attached to phosphatidic acids or to 1,2-
   diacylglycerols are made available in an activated form, usually by
   association with a nucleotide such as cytidine or uridine phosphate.
5. Acylation is usually effected by fatty acids as their coenzyme A thiol
   esters.
6. In plants these compounds are produced entirely by biosynthesis but in
   animals full biosynthesis is supplemented by modification of dietary
   lipids.

The key intermediates in lipid biosynthesis are phosphatidic acids,
produced mainly from sn-glycerol-3-phosphate but also, to a minor degree,
from dihydroxyacetone and the 1,2-diacylglycerols produced from phosphatidic acids by dephosphorylation. Further acylation of the diacylglycerols
with acyl-CoA furnishes the triacylglycerols. An alternative route to
diacylglycerols in animals is acylation of the 2-monoacylglycerols produced
by lipolysis of dietary triacylglycerols.

Phosphatidylinositides and phosphatidylglycerols are produced from
phosphatidic acids via CDP-diacylglycerols. Phosphatidylcholines and
phosphatidylethanolamines are derived from diacylglycerols by reaction
with CDP-choline and CDP-ethanolamine, respectively. The phosphatidylethanolamines act as a source of phosphatidylserines by base exchange and
can also furnish phosphatidylcholines by stepwise methylation. Reaction of
diacylglycerols wtih UDP-galactose gives first the monogalactosyl- and
then the digalactosyldiacylglycerols. These changes are formulated in
Figure 2.1.

Sphingenine (and related long-chain bases) are made from palmitoyl-
CoA and serine by the route set out in Figure 2.2. This, in turn, is
converted to cerebrosides or to sphingomyelin, in each case by two
alternative routes (Figures 2.3 and 2.4).

2.9   Chemical synthesis of acylglycerols and phospholipids

2.9.1   Introduction

The position of acyl groups linked to glycerol will be shown by numbers 1,
2, and 3 and the compounds are racemic unless otherwise indicated by the
prefix sn-.

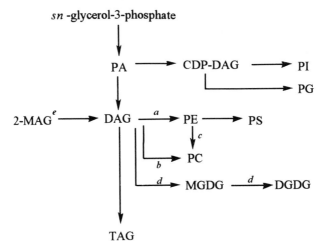

Figure 2.1 Biosynthesis of glycerol esters (PA = phosphatidic acids, PE = phosphatidyl-ethanolamines, PC = phosphatidylcholines, PI = phosphatidylinositides, PS = phosphatidyl-serines, PG = phosphatidylglycerols, MAG = monoacylglycerols, DAG = diacylglycerols, TAG = triacylglycerols, UDP = uridine diphosphate, CDP = cytidine diphosphate). *a*, reaction with CDP–ethanolamine. *b*, reaction with CDP–choline. *c*, stepwise methylation with *S*-adenosylmethionine. *d*, UDP-galactose. *e*, from dietary TAG.

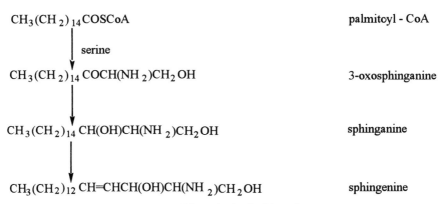

Figure 2.2 Biosynthesis of sphingenine.

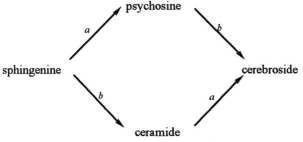

Figure 2.3 Biosynthesis of cerebrosides. *a*, UDP-galactase. *b*, acyl-CoA.

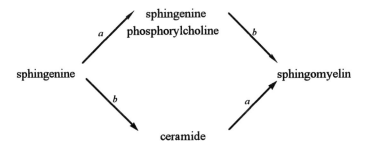

Figure 2.4 Biosynthesis of sphingomyelin. *a*, CDP–choline. *b*, acyl-CoA.

A major problem in the synthesis of pure acyl glycerols is the ease with which acyl groups in mono- and diacylglycerols migrate from the oxygen to which they are attached to an adjacent free hydroxyl group. This transesterification process is catalysed by acid or base or heat and may even occur during chromatography. The 1- and 2-monoacylglycerols readily furnish a 90:10 mixture of these two isomers and 1,2- and 1,3-diacylglycerols form mixtures with 60–80% of the latter.

$$
\begin{bmatrix} OCOR \\ OH \\ OH \end{bmatrix} \rightleftharpoons \begin{bmatrix} O \\ O \\ OH \end{bmatrix}\!C\!\begin{matrix} OH \\ OR \end{matrix} \rightleftharpoons \begin{bmatrix} OH \\ OCOR \\ OH \end{bmatrix}
$$

Procedures for obtaining acylglycerols with the desired acyl groups attached at the appropriate positions are based on the use and removal of protecting or blocking groups or on the greater reactivity of primary over secondary hydroxyl groups. It is essential when removing protecting groups to avoid conditions which promote acyl migration.

Purification of intermediates and final products is effected by careful crystallization or by chromatography. The purity of synthetic glycerides is checked by gas chromatography, thin-layer chromatography, high-performance liquid chromatography, and by enzymic and spectroscopic procedures.

2.9.2 Acylation

The following acylation procedures may be employed:

(i) Reaction at room temperature for 1–3 days or at 100°C for 4 hours with a slight excess of acyl halide and an equivalent amount of pyridine or other tertiary base in chloroform solution. Acid chlorides are prepared from the appropriate acid using excess of thionyl chloride ($SOCl_2$) for saturated acids or of oxalyl chloride [$(COCl)_2$] for unsaturated acids.

(ii) As in (i) but replacing acyl halide with acid anhydride and trifluoromethylsulphonic acid (CF_3SO_3H).

(iii) Direct reaction with a carboxylic acid and an appropriate catalyst such as *p*-toluenesulphonic acid, trifluoroacetic anhydride, or sulphonated polystyrene resin. Acylation with free acid in the presence of 1,1'-dicyclohexylcarbodi-imide and 4-dimethylaminopyridine (DMAP) occurs at room temperature in 2 hours in tetrachloromethane solution.

2.9.3   Protecting groups

The selection of appropriate protecting groups, their use in an appropriate order, and their removal under suitably mild conditions is the key to the successful synthesis of acylglycerols. Some of those most commonly employed are listed in Table 2.7.

Glycerol reacts with acetone and an acidic catalyst to give 1,2-isopropylidine glycerol leaving one free primary hydroxyl group which can be acylated or alkylated followed by the removal of the isopropylidene group under conditions such that acyl migration does not occur. Benzaldehyde reacts similarly to give the 1,2- and 1,3- acetals which can be separated by crystallization. 1,3-Benzylidene glycerol has a free secondary hydroxyl group which can be acylated.

1,2-isopropylidine   1,3- and 1,2-benzylidene glycerol
glycerol

Other protecting groups based on ethers (benzyl, trityl) or esters are given in Table 2.7 and are used in some of the synthetic sequences described later.

2.9.4   C₃ compounds other than glycerol

As an alternative to the use of protecting groups glycerol may be replaced by other C_3 compounds which allow a stepwise reaction at each carbon atom. Examples include the following:

$HOCH_2COCH_2OH$	dihydroxyacetone	1,3-dihydroxypropan-2-one
$CH_2{=}CHCH_2OH$	allyl alcohol	propenol
$CH_2 CHCH_2OH$ (epoxide)	glycidol	2,3-epoxypropanol
$ClCH_2CH(OH)CH_2OH$	-	3-chloropropane-1,2-diol

Table 2.7 Protecting groups in the formation of acylglycerols

Protecting group	Reagent for formation	Reagent for removal
isopropylidine	acetone, p-TSA[a]	aqueous acid, $B(OMe)_3$ and H_3BO_3, $B(OEt)_3$ followed by H_2O, Me_2BBr, $-50°C$
benzylidene	benzaldehyde, p-TSA[a]	hydrogenolysis[b]
benzyl	benzyl chloride	hydrogenolysis[b]
4-methylbenzyl	4-methylbenzyl chloride	DDQ[c]
2-iodobenzyl	2-iodobenzyl chloride	chlorination, $MeOH$, Na_2CO_3 or $NaHCO_3$
trityl[d]	Ph_3CCl[e]	hydrogenolysis,[b] HCl in Et_2O or pet. ether, HBr in AcOH, BF_3–$MeOH$–CH_2Cl_2,
trichloroethoxycarbonyl	CCl_3CH_2OCOCl	Zn, AcOH
carbonate	phosgene	mild alkaline hydrolysis

[a]p-toluenesulphonic acid, [b]H_2, catalyst; [c]2,3-dichloro-5,6-dicyano-1,4-benzoquinone, [d]triphenylmethyl (Ph_3C), [e]reacts with primary alcohol groups.

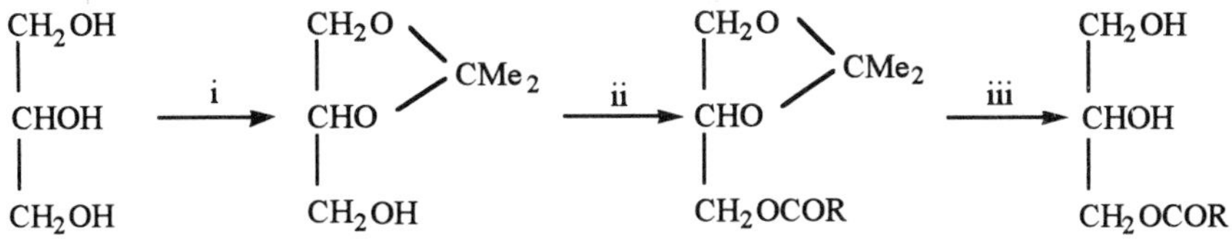

Figure 2.5 Preparation of 1-monoacylglycerols. Reagents: i, $COMe_2$, C_6H_6, p-MeC$_6$H$_4$SO$_3$H; ii, RCO_2H, p-MeC$_6$H$_4$SO$_3$H; iii, H_3BO_3, $B(OMe)_3$; H_2O.

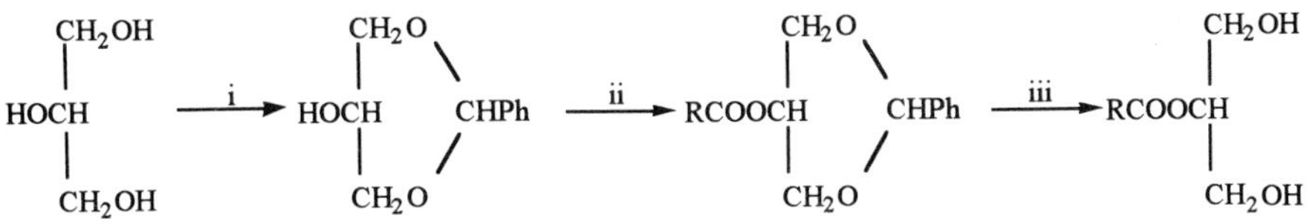

Figure 2.6 Preparation of 2-monoacylglycerols. Reagents: i, PhCHO, p-MeC$_6$H$_4$SO$_3$H; ii, RCOCl, pyridine; iii, H_3BO_3, $B(OMe)_3$; H_2O.

2.9.5 Synthetic procedures for racemic acylglycerols

Some typical examples of the synthesis of mono-, di-, and triacylglycerols are given in Figures 2.5–2.10. These are representative of the large number of synthetic sequences which have been reported. Some of these sequences can be shortened insofar as it is now possible to purchase some of the starting materials and intermediates.

Monoacylglycerols are usually prepared from 1,2-isopropylidene glycerol (Figure 2.5) or from 1,3-benzylidene glycerol (Figure 2.6).

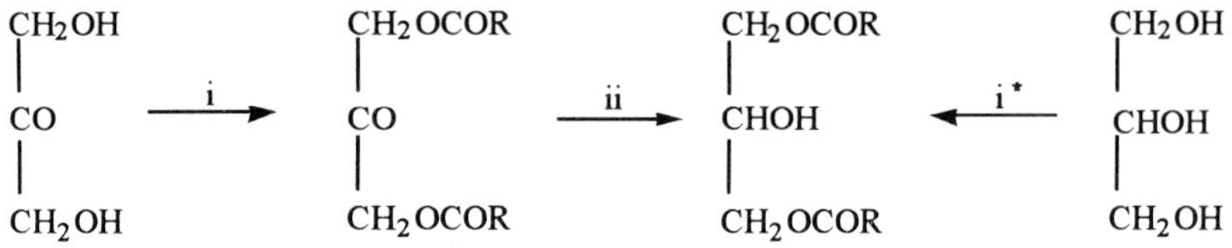

CH₂OH   CH₂OCOR   CH₂OCOR   CH₂OH

The scheme shows:

$$CH_2OH | CO | CH_2OH \xrightarrow{i} CH_2OCOR | CO | CH_2OCOR \xrightarrow{ii} CH_2OCOR | CHOH | CH_2OCOR \xleftarrow{i^*} CH_2OH | CHOH | CH_2OH$$

Figure 2.7 Synthesis of 1,3-diacylglycerols with identical acyl groups. Reagents: i, RCOCl, pyridine; ii, NaBH₄.
*primary alcohol groups are more reactive than secondary alcohol groups but the product will not be pure.

$$CH_2OCOR^1 | CHOH | CH_2OH \longrightarrow CH_2OCOR^1 | CHOH | CH_2OCOR^2$$

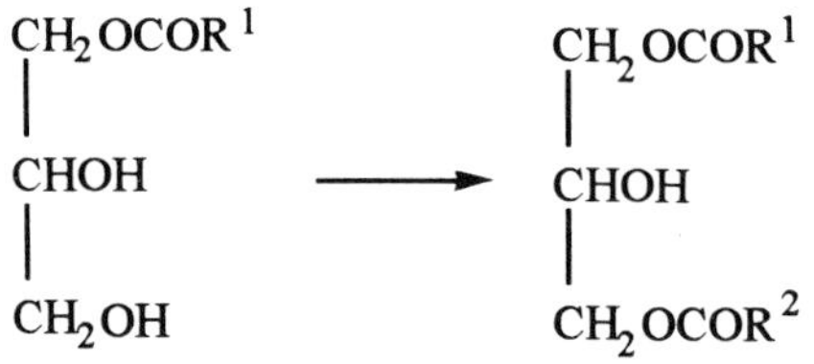

Figure 2.8 Synthesis of 1,3-diacylglycerols with different acyl groups. Reagents: R²COCl, pyridine (primary alcohol groups are more reactive than secondary alcohol groups but the product will not be pure).

1,3-diacylglycerols can be prepared in several ways depending on whether the acyl groups are the same (Figure 2.7) or different (Figure 2.8). In the latter case reaction proceeds *via* the appropriate monoacylglycerol. 1,2-diacylglycerols are more difficult to make because of the readiness with which they undergo acyl migration and because they are not so easily crystallized. Some examples are given in Figure 2.9. These make use of four different starting materials. The only new feature in these is in route 2 where an enzyme (pancreatic lipase) is used to selectively remove an acyl group from the α-position.

The equilibrium which exists between 1,2- and 1,3-diacylglycerols and the lower solubility of the latter has been exploited in synthesis. Crystallization of a mixture of the two diacylglycerols from a suitable solvent such as hexane gives the 1,3- isomer only.

Triacylglycerols containing only one type of acyl group are easily made by acylation of glycerol with acid chloride, anhydride, or free acid. For example: glycerol and a small excess of fatty acid with *p*-toluenesulphonic acid as catalyst, tetrahydrofuran as solvent, and molecular sieve as desiccant give good yields of triacylglycerols after reaction at 110–120°C for about 6 hours. Those containing two or three different acyl groups are generally made by extension of the procedures employed for monoacyl-glycerols and diacylglycerols. In designing a synthetic route the aim should

Route 1

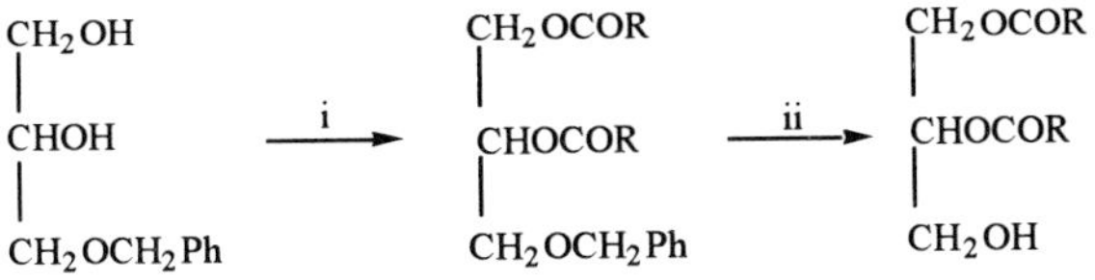

Route 2

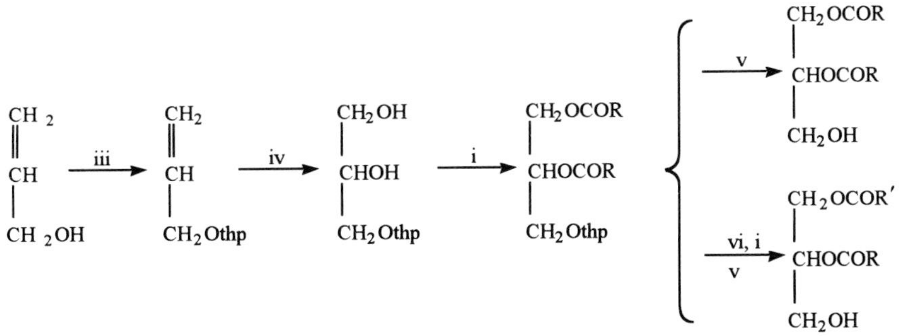

Routes 3 and 4

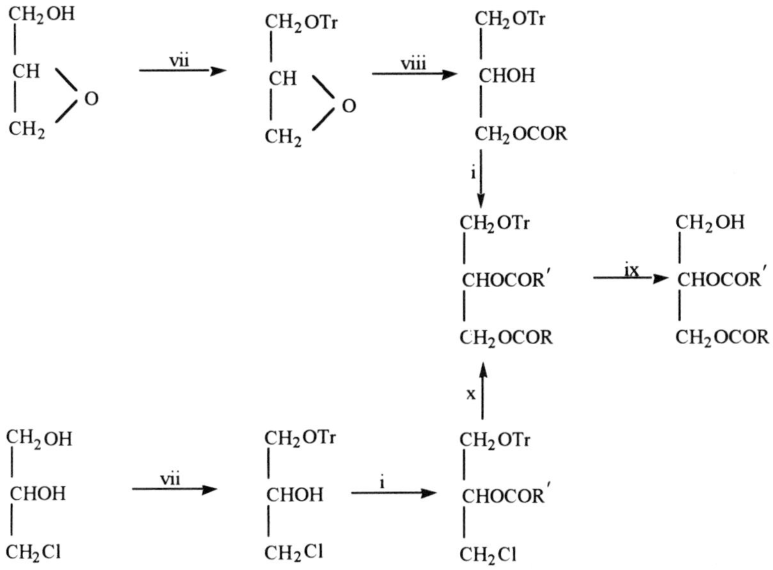

Figure 2.9 Preparation of 1,2-diacylglycerols (routes 1–4). Reagents: i, acyl halide, pyridine; ii, Ni, H$_2$; iii, dihydropyran, H$^+$; iv, KMnO$_4$; v, HCl or B(OH)$_3$; vi, pancreatic lipase; vii, Ph$_3$CCl; viii, RCOOH, ix, (CF$_3$CO)$_2$O, MeOH; x, RCOONa (thp = tetrahydropyranyl, Tr = triphenylmethyl).

be to proceed through the most stable intermediates (avoid 1,2-diacyl-glycerols) and introduce unsaturated acyl groups as late as possible. Some examples are given in Figure 2.10.

2.9.6 Synthetic procedures for enantiomeric acylglycerols

The procedures described in the preceding sections yield racemic products. Enantiomeric glycerol esters are prepared by modification of these pathways using chiral intermediates.

1,2-isopropylidene-sn-glycerol can be prepared from natural (D)-mannitol (Figure 2.11). 2,3-isopropylidene-sn-glycerol is more difficult to obtain since (L)-mannitol has to be synthesized from L-arabinose. Alternatively, the 1,2-ketal can be transformed to the 2,3-ketal by a series of stereospecific reactions. A different approach starts with serine. The L-isomer furnishes 2,3-isopropylidene-sn-glycerol from which 1-benzyl-sn-glycerol can be obtained: the D-isomer yields 1,2-isopropylidine-sn-glycerol and 3-benzyl-sn-glycerol (Figure 2.12).

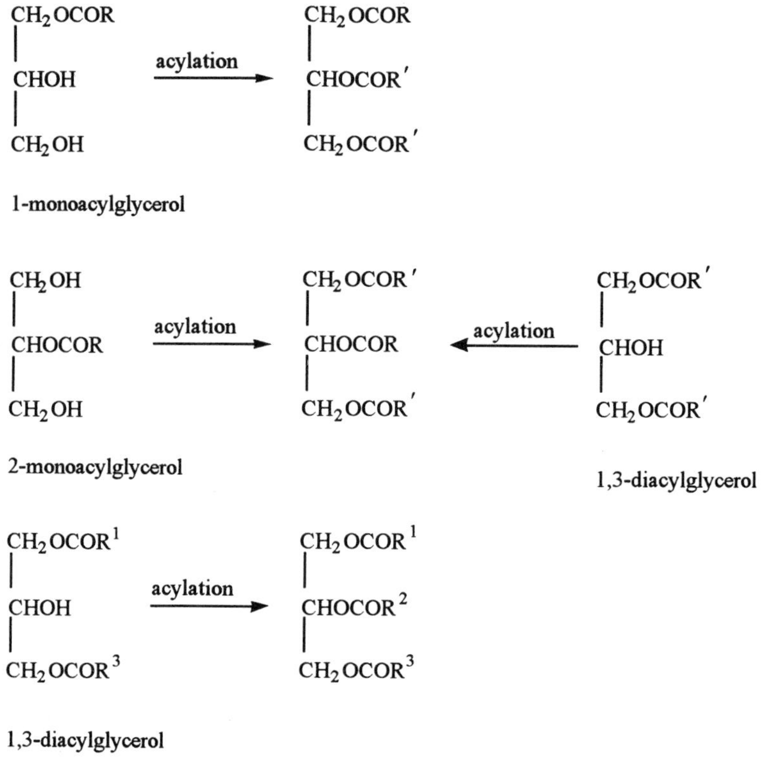

$$CH_2OCOR \quad \xrightarrow{\text{acylation}} \quad CH_2OCOR$$

CH_2OCOR	CH_2OCOR
$CHOH$ $\xrightarrow{\text{acylation}}$	$CHOCOR'$
CH_2OH	CH_2OCOR'

1-monoacylglycerol

CH_2OH	CH_2OCOR'	CH_2OCOR'
$CHOCOR$ $\xrightarrow{\text{acylation}}$	$CHOCOR$ $\xleftarrow{\text{acylation}}$	$CHOH$
CH_2OH	CH_2OCOR'	CH_2OCOR'

2-monoacylglycerol   1,3-diacylglycerol

CH_2OCOR^1	CH_2OCOR^1
$CHOH$ $\xrightarrow{\text{acylation}}$	$CHOCOR^2$
CH_2OCOR^3	CH_2OCOR^3

1,3-diacylglycerol

Figure 2.10 Preparation of triacylglycerols.

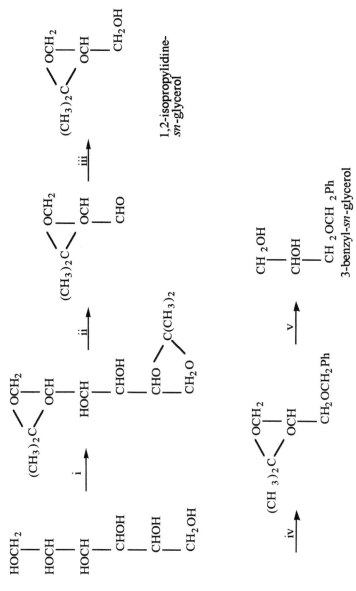

Figure 2.11 Preparation of 1,2-isopropylidene-*sn*-glycerol and 3-benzyl-*sn*-glycerol from D-mannitol. Reagents: i, CO(CH₃)₂, ZnCl₂; ii, Pb(OAc)₄; iii, LiAlH₄; iv, Na, PhCH₂OH; v, B(OH)₃.

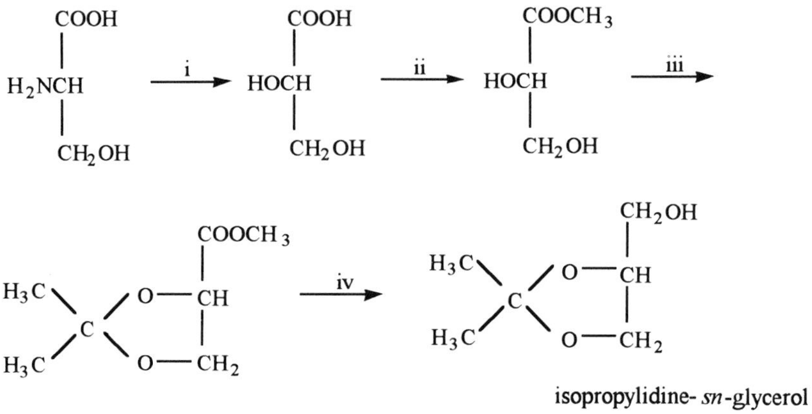

Figure 2.12 Synthesis of 1,2- and 2,3-isopropylidene-*sn*-glycerol from D- and L-serine, respectively. Reagents: i, NaNO$_2$, HCl; ii, CH$_3$OH; iii, CO(CH$_3$)$_2$, (CH$_3$)$_2$C(OCH$_3$)$_2$; iv, LiAlH$_4$.

Yet other syntheses are based on glycidol which is available in R- and S-forms (Figures 2.13 and 2.14). The synthesis of 1,2-diacyl-*sn*-glycerols are of special interest because such compounds are required to synthesize enantiomeric phospholipids.

2.9.7 Structured lipids

Structured lipids have been defined as triacylglycerols that have been altered by some process in order to meet more closely a specification expressed in terms of fatty acid or, more usually, triacylglycerol composition. The synthetic methods which have been described in the previous sections are suitable on a laboratory scale but are not economic on a commercial scale even when the product has enhanced value. Here the objective is to obtain a product as close as possible to the specification at an appropriate cost.

Structural lipids are designed to modify one or more of the physical (melting behaviour, polymorphic nature), chemical (oxidative stability) or nutritional properties (presence or absence of saturated acids or of particular essential fatty acids, ease of absorption and digestion) of the oils.

The ways of modifying natural oils and fats to provide more valuable products include:

- blending,
- fractionation,
- interesterification,
- partial hydrogenation,

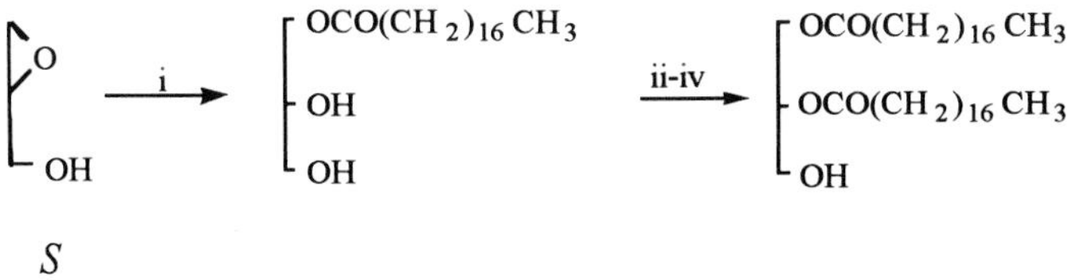

$$\begin{array}{c}\overset{\bullet}{\underset{\text{OH}}{\biggr[}}\overset{\text{O}}{}\end{array} \xrightarrow{\text{i}} \left[\begin{array}{l}\text{OCO(CH}_2)_{16}\,\text{CH}_3 \\ \text{OH} \\ \text{OH}\end{array}\right. \xrightarrow{\text{ii-iv}} \left[\begin{array}{l}\text{OCO(CH}_2)_{16}\,\text{CH}_3 \\ \text{OCO(CH}_2)_{16}\,\text{CH}_3 \\ \text{OH}\end{array}\right.$$

S

Figure 2.13 Synthesis of 1-monoacyl and 1,2-diacyl-*sn*-glycerols from glycidol (the 1,2-diacylglycerol can be used to prepare enantiomeric triacylglycerols or phospholipids). Reagents: i, stearic acid, Ti(OPr^{i})$_4$; ii, Me$_2$Bu^{t}SiCl, imidazole (to block *sn*-3 position); iii, stearoyl chloride, pyridine; iv, NBS, DMSO, H$_2$O, THF (to remove blocking group). (Burgos (1987)).

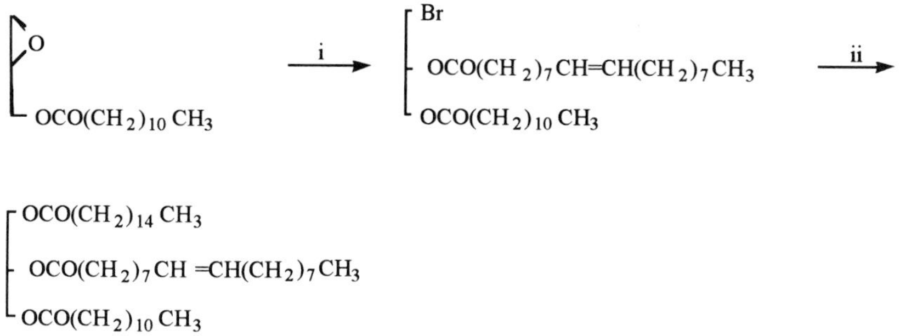

$$\left[\begin{array}{l}\overset{\text{O}}{} \\ \text{OCO(CH}_2)_{10}\,\text{CH}_3\end{array}\right. \xrightarrow{\text{i}} \left[\begin{array}{l}\text{Br} \\ \text{OCO(CH}_2)_7\text{CH=CH(CH}_2)_7\text{CH}_3 \\ \text{OCO(CH}_2)_{10}\,\text{CH}_3\end{array}\right. \xrightarrow{\text{ii}}$$

$$\left[\begin{array}{l}\text{OCO(CH}_2)_{14}\,\text{CH}_3 \\ \text{OCO(CH}_2)_7\text{CH=CH(CH}_2)_7\text{CH}_3 \\ \text{OCO(CH}_2)_{10}\,\text{CH}_3\end{array}\right.$$

Figure 2.14 Synthesis of enantiomeric triacylglycerols from *R*-glycidol (the enantiomer is produced by starting with the palmitate in place of the laurate). Reagents: i, oleic anhydride, LiBr, THF or benzene; ii, palmitic acid (as salt), THF, HMPT, 45–50°. (Sonnet (1991); Sonnet and Dudley (1994)).

- modification of seed lipids by plant breeding or application of gene technology,
- glyceride synthesis by chemical or enzymic procedures.

Only the last of these will be discussed here. Some of the others are discussed in sections 4.3–4.5. Some examples are described here and in section 8.3.

(i) Glycerol 1,3-dibehenate-2-oleate (BOB) inhibits fat bloom when added to chocolate. It can be produced by enzymic interesterification as, for example, by reaction of triolein with behenic acid or ester in the presence of a 1,3-stereospecific lipase.

(ii) Betapol* (glycerol 1,3-dioleate-2-palmitate, OPO). This symmetrical dioleopalmitin with palmitic acid in the β-position, though generally absent from seed oils, is an important constituent of human milk fat. It is therefore a desirable constituent of infant formula and can be made from glycerol tripalmitate, oleic acid or ester, and a 1,3-stereospecific lipase.

---

* Betapol, Captrin, Caprenin and Olestra are trade names associated with products of the type described here and each made by one company.

(iii) Oils rich in lauric and linoleic acid can be made by interesterification of mixtures of oils rich in these acids.

(iv) Acids or esters with unsaturation at $\Delta 4$ (e.g. DHA), $\Delta 5$ (e.g. EPA), or $\Delta 6$ (e.g. GLA) react more slowly in enzymic processes than saturated acids or those having $\Delta 9$ unsaturation and this property has been exploited in several ways to increase the concentration of these acids. For example, tuna oil with about 25% of DHA when subjected to lipolysis gives a mixture of triacylglycerols and diacylglycerols together containing about 53% of DHA because these glycerol esters are more resistant to lipolysis than the esters which do not have a $\Delta 4$ double bond.

(v) Medium chain triglycerides (MCT) contain virtually only C_8 and C_{10} acids derived from minor components of coconut oil or palmkernel oil. They are made by catalysed reaction between glycerol and these medium-chain acids. MCT contain no unsaturated acids and are oxidatively stable. This, combined with their liquid nature and their viscosity, makes them suitable as lubricants and release agents and they are much used for this purpose in the food industry. Accidental contamination of food product with this edible lubricant would not create any hazard. The ease with which MCT are absorbed and metabolized makes them important components of foods devised for those with special needs such as athletes, invalids, and babies. One such preparation is designated Captrin.*

(vi) Caprenin* is a semi-synthetic glycerol triester containing approximately equimolar proportions of octanoic (8:0), decanoic (10:0), and docosanoic acids (22:0) with some of the latter replaced by small amounts of 20:0 and 24:0. Since the three major acids differ so much in molecular weight the composition in terms of percentage weight is somewhat different. An equimolar mixture of the three major components would correspond to 22.6% C_8 26.7% C_{10} and 50.7% C_{22} by weight.

The melting behaviour of Caprenin is similar to that of cocoa butter and it can be used in soft candy and in confectionary coating for nuts and fruits etc. The medium-chain acids are metabolized in the usual way but behenic acid (22:0) is poorly absorbed so that around 86% of this high-melting long-chain acid is excreted. As a consequence of this the oil has a calorific value of little over one half that of usual fats (5 as opposed to 9 kcal/g).

In one method of preparation glycerol behenates (mono, di, and tri) are esterified with capric acid and the resulting triacylglycerol is interesterified (NaOH at 80°C) with MCT containing appropriate levels of caprylic and capric esters. The product contains around 92% of mono-long-chain triacylglycerols with C_{38} (around 22%), C_{40} (about 48%), and C_{42} (around 24%) compounds as the major components.

(vii) Salatrim contains triacylglycerols of short-chain acids with hydrogen-

---

* See footnote on previous page.

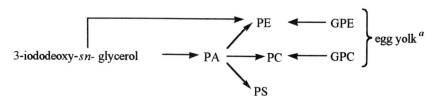

Figure 2.15 Summary of synthetic routes to phospholipids (PA = phosphatidic acids, PE = phosphatidylethanolamines, PC = phosphatidylcholines, PS = phosphatidylserines, GPE = glycerylphosphoethanolamines, GPC = glycerophosphocholines). *a*, or other natural source of pure phospholipid.

ated canola, soybean, or cottonseed oils with intended use in chocolate, dairy products, and salty snacks.

(viii) Olestra* is a sucrose derivative with six–eight esterified groups obtained from sucrose and common vegetable oils (soybean, corn, cottonseed, sunflower, rapeseed). It is prepared by reaction of high-grade sucrose with high-grade methyl esters in the presence of sodium or potassium soaps to improve solubility and an alkali carbonate as catalyst. It can be used as a frying oil and can replace fats in products such as ice-cream, margarine, cheese, and baked goods. The material is non-toxic, non-carcinogenic, and is so poorly absorbed as to have zero caloric value. Permission has not yet been given for its use in food.

2.9.8 Synthesis of phospholipids

This brief account is restricted to the more important methods of preparing phosphatidyl esters. Further details and further references are available in *The Lipid Handbook*. These compounds generally have four ester linkages and the major preparative procedures vary in the order in which these are assembled. The most common starting materials are:

1. 1,2-diacylglycerols or the related iodo compound;
2. glycerophosphocholine or glycerophosphoethanolamine obtained from natural sources;
3. phosphatidic acids.

The relationships between these are summarized in Figure 2.15 and some of these are elaborated in the following paragraphs.

Optically active compounds are prepared from enantiomeric intermediates prepared as already described or obtained from natural sources by enzymic reactions. A simple way of producing phosphatidyl esters with two different acyl groups involves selective enzymic deacylation and chemical reacylation with acyl chloride or anhydride.

* See footnote on page 55.

The conversion of 1,2-diacylglycerol to the iodide and its subsequent conversion to phosphatidylcholines or phosphatidylethanolamines with appropriate silver salts is set out in Figure 2.16.

Pure PC (or PE) can be obtained from egg yolk or from some other natural source. These are enantiomeric compounds retaining the stereochemistry of the natural materials. They are, however, mixtures of compounds because of the different acyl chains that they contain. The acyl chains can be removed without loss of optical purity and the resulting GPC (or GPE) can be reacylated to give PC (PE) compounds with two identical acyl groups.

If a phospholipid with two different acyl groups is the target of synthesis then the acyl chain on C2 can be selectively removed by enzymic deacylation with phospholipase A2 and the lyso-PC reacylated with the appropriate acyl anhydride (Figure 2.17).

Phosphatidic acids, prepared from 1,2-diacylglycerols or the corresponding iodide, can be converted to phosphatidylethanolamimes and phosphatidylcholines by reaction with tri-isopropylbenzenesulphonyl chloride to give the acid chloride which is then reacted with an appropriate choline or ethanolamine derivative (Figure 2.18).

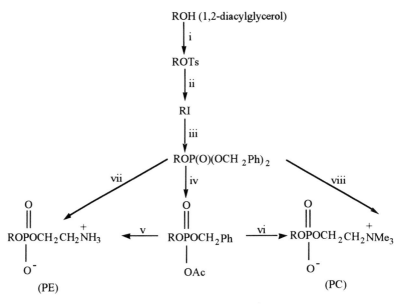

Figure 2.16 Synthesis of phospholipids from 1,2-diacylglycerols *via* the 3-iododeoxy derivative. Reagents: i, toluenesulphonyl chloride, TsCl; ii, NaI; iii, AgOP(O)(OCH$_2$Ph)$_2$; iv, NaI, AgNO$_3$; v, BrCH$_2$CH$_2$N(CH$_2$Ph)$_2$, H$_2$/Pd; vi, BrCH$_2$CH$_2$N Me$_3$ picrate, NaI; vii, AgOP(OY)CH$_2$CH$_2$NHX (Y = Bu^t or Ph or PhCH$_2$, X = (CO)$_2$C$_6$H$_4$ or COOCH$_2$Ph or CPh$_3$); viii, AgOP(O)(OCH$_2$C$_6$H$_4$NO$_2$-p) OCH$_2$CH$_2$Cl; NMe$_3$.

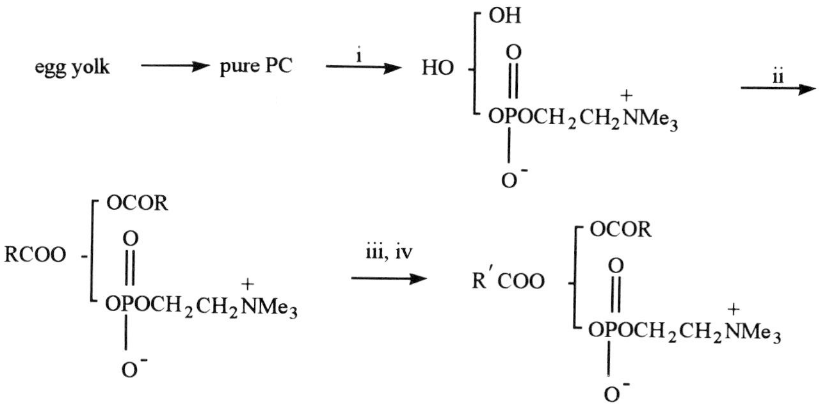

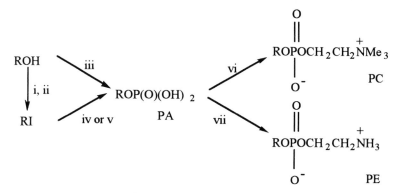

Figure 2.17 Preparation of phosphatidylcholines from natural glycerophosphocholines. Reagents: i, Bu₄N ŌH; ii, (RCO)₂O, RCOOK; iii, phospholipase A2; iv, (R′CO)₂O, R′COOK.

Figure 2.18 Synthesis of phosphatidic acids and esters (ROH = 1,2-diacylglycerol). Reagents: i, TsCl; ii, NaI; iii, POCl₃; H₂O; iv, AgQPO(OCH₂Ph)₂, BaI; v, AgOPO(OBuᵗ)₂, HCl; vi, 2,4,6-(Prⁱ)₃C₆H₂SO₂Cl, HŌCH₂CH₂N Me₃ŌAc; vii, 2,4,6-(Prⁱ)₃C₆H₂SO₂Cl, HOCH₂CH₂NHCPh₃, H₂Pd.

Some of the less common phosphatidyl lipids have been prepared on a commercial scale from chromatographically purified natural PC or PE products by exchange of the choline or ethanolamine moieties with other units under the influence of a phospholipase. (This is another example of the production of a structured lipid.)

Bibliography

Blomquist, G.J. and L.L. Jackson (1979) Chemistry and biochemistry of insect waxes, *Prog. Lipid Res.*, **17**, 319–345.

Burgos, C.E., D.E. Ayer and R.A. Johnson (1987) *J. Org. Chem.*, **56**, 4973–4977.

Cevc, G. (ed.) (1993) *Phospholipids Handbook*, Dekker, New York.

Gurr, M.I. and J.L. Harwood (1991) *Lipid Biochemistry – an Introduction*, 4th edn, Chapman and Hall, London.

Hamilton, R.J. (ed.) (1995) *Waxes: Chemistry, Molecular Biology and Functions*, The Oily Press, Dundee.

Harwood, J.L. (1994) in *The Lipid Handbook*, 2nd edn (eds F.D. Gunstone, J.L. Harwood, and F.B. Padley), Chapman and Hall, London, Chapter 2.

Harwood, J.L. and J.R. Bowyr (eds) (1990) *Lipids, Membranes, and Aspects of Photobiology*, Volume 4 of *Methods in Plant Biochemistry*, Academic Press, London.

Harwood, J.L. and N.J. Russell (1984) *Lipids in Plants and Microbes*, George Allen and Unwin, London.

Kates, M. (ed.) (1990) *Handbook of Lipid Research*, Volume 6, *Glycolipids, Phosphoglycerides, and Sulfoglycolipids*, Plenum, New York.

Kolattukudy, P.E. (ed.) (1976) *Chemistry and Biochemistry of Natural Waxes*, Elsevier, Amsterdam.

Murphy, D.J. (ed.) (1993) *Designer Oil Crops: Breeding, Processing, and Biotechnology*, VCH, Weinheim.

Puleo, S. and T.P. Rit (1992) Natural waxes: past, present, and future, *Lipid Technol.*, **4**, 82–90.

Sonnet, P.E. (1991) *Chem. Phys. Lipids*, **58**, 35–39.

Sonnet, P.E. and R.L. Dudley (1994) *Chem. Phys. Lipids*, **72**, 185–191.

Szuhaj, B.F. and G.R. List (1985) *Lecithins*, AOCS.

3 The major sources of oils, fats, and other lipids

3.1 Fats and oils in the marketplace

The annual production of oils and fats was 83.7 million tonnes (mt) in the 1991–92 harvest year of which about 23.7 mt (31%) entered into world exports and the remainder was used in the country where it was produced. Three-quarters of the total production is of vegetable origin. Most of this material (around 80%) is used for human food and a further 6% is used for animal feed to produce yet more human food. The remainder (about 14%, equivalent to almost 12 mt) serves as source material for the oleochemical industry. About 90% of this is used in the production of soap and other surface-active compounds and the balance, representing about 1.2 mt, is used for other industrial purposes.

Annual production is expected to rise from 84 mt in 1991–92 to over 100 mt by the year 2000. These figures can be compared with the annual production of mineral oil and coal (each about 3000 mt), of wheat and rice (each about 400 mt), and of potatoes (about 300 mt). The annual production of sugar is comparable with that of oils and fats.

Oils and fats come from vegetable and animal sources and can be further categorized as in Table 3.1. Soybean and cottonseed oils, of which the former is particularly important, are by-products of plants grown primarily for their seed meal (a protein-rich animal feed) and fibre respectively. Rape, sunflower, and groundnut, on the other hand, are grown as oil crops with the residual meal as a by-product. These are annual crops and the quantities cultivated can be changed from one season to the next depending on the perceived financial return to the grower compared with that for alternative crops. In contrast, oil-bearing trees such as olive, palm, and coconut, once mature, continue to bear a useful harvest for many years. Linseed, castor, and tung are vegetable oils produced only for non-food uses (but see linola oil, section 3.7.5). Animal fats come from cattle, sheep, and pigs or are of marine origin.

Table 3.2a shows the world production of these major oils from 1965 to the present with forecasts for 1995 and 2005. Over this 40-year period total oil/fat production is expected to rise almost four-fold from 32 mt in 1965 to 115 mt in 2005. All the materials listed in this table show increasing production over this time scale. Some have kept pace with the total growth while others have increased or decreased their market share. The most significant oils within this range are the products of the oil palm (palm oil

Table 3.1 Major sources of oils and fats

Plant (vegetable) sources	Examples
vegetable oils – by-products	soybean, cottonseed, rice bran, corn
vegetable oils – primary products	rape, sunflower, groundnut
tree crops	olive, palm, palmkernel, coconut
industrial oils	linseed, castor, tung, tall oil[a]
Animal sources	
land animals	milk fat, tallow, lard
marine animals	fish

[a]Tall oil is a product of the wood pulp industry and consists of fatty acids and resin acids. It could be considered as a by-product or as a tree crop.

and palmkernel oil), coconut oil, and oils from soybean, rape, and sunflower. The proportion of these six oils represented 33% of total oil and fat production in 1965, rose to 55% by 1985, and is expected to be 63% in 2005. They make up an even higher proportion of oils which are traded across national boundaries. The four C_{16}/C_{18} oils and the two lauric oils (palmkernel and coconut) show the greatest growth in the oils and fats sector. More recent forecasts for the quinquennia 1993–97 and 2008–12 collected in Table 3.2b show the continued dominance of these oils.

The figures in Table 3.3 illustrate further the very large increase in the availability of palm oil (accompanied inevitably by palmkernel oil). This development has occurred particularly in Malaysia and Indonesia, although it is not confined to these two south-eastern Asian countries. Still further rises in these production figures are expected from existing and from new plantings.

The demand for oils and fats for food and non-food use is linked particularly with growing world population and with income growth. Up to a certain level of income, demand for oils and fats increases with growing prosperity. On the other hand, the supply of these materials from vegetable sources is linked with the area of land under appropriate cultivation and with increasing productivity. The figures in Table 3.4 show how much greater is the production of oil in the tropical areas.

In 1991–92 the average annual usage of fats for both food and non-food uses was 15.3 kg per person. This covers a range of values from 35–40 kg in most affluent countries (North America, Western Europe, and Australasia), through about 7 kg in the highly-populated countries of India and China, to even lower levels in many of the poorer countries. In view of the very large populations in India and China the global demand for oils and fats is likely to grow for some time yet. The general relation between fat consumption and income does not apply in quite the same way in Japan where lower-than-expected consumption is linked to different dietary customs.

It is worth noting that while those living in the affluent countries are

Table 3.2a Major vegetable and animal oils and fats – changes in world production over the period 1965–2005

Oil/fat	Millions of tonnes (% of total)				
	1965	1975	1985	1995[a]	2005[a]
soya	4.1 (13.0)	8.5 (19.7)	14.1 (22.1)	19.5 (22.1)	27.3 (23.8)
palm	1.4 (4.4)	2.8 (6.5)	6.7 (10.5)	13.9 (15.7)	21.4 (18.7)
rape	1.4 (4.4)	2.6 (6.1)	6.0 (9.4)	9.5 (10.7)	12.0 (10.5)
sunflower	2.9 (9.2)	3.7 (8.6)	6.5 (10.2)	8.9 (10.0)	10.8 (9.4)
cotton	2.6 (8.3)	2.9 (6.8)	3.4 (5.3)	4.4 (5.0)	5.2 (4.5)
groundnut	3.0 (9.5)	2.9 (6.8)	3.3 (5.2)	4.2 (4.7)	5.1 (4.4)
coconut	2.0 (6.3)	2.6 (6.1)	2.7 (4.2)	3.2 (3.6)	3.5 (3.1)
palmkernel	0.4 (1.3)	0.5 (1.2)	0.9 (1.4)	1.7 (1.9)	2.7 (2.4)
corn	0.4 (1.3)	0.6 (1.4)	1.0 (1.6)	1.6 (1.8)	2.0 (1.7)
subtotal	18.2 (57.7)	27.1 (63.2)	44.6 (69.9)	66.9 (75.5)	90.0 (78.5)
butter fat	4.6 (14.6)	5.3 (12.4)	6.3 (9.9)	6.7 (7.6)	7.3 (6.4)
tallow	4.3 (13.7)	5.5 (12.8)	6.4 (10.0)	7.3 (8.2)	8.1 (7.1)
lard	3.5 (11.1)	4.0 (9.3)	5.0 (7.8)	6.2 (7.0)	7.5 (6.6)
fish	0.9 (2.9)	1.0 (2.3)	1.5 (2.4)	1.5 (1.7)	1.6 (1.4)
subtotal	13.3 (42.3)	15.8 (36.8)	19.2 (30.1)	21.7 (24.5)	24.5 (21.5)
total	31.5	42.9	63.8	88.6	114.5

[a]Forecast
Source: *Oil World 1958–2007* (1988), ISTA Mielke GmbH, Hamburg.

Table 3.2b Oil usage forecast (millions of metric tonnes)

Oil/fat	1993–97 average	2008–12 average	Percentage increase[a]
palm	15.4	29.8	94
soybean	17.9	25.1	41
rapeseed	10.1	15.6	54
sunflower	8.4	12.0	44
tallow/grease	7.0	8.1	16
lard	5.7	7.7	35
butter (as fat)	6.0	6.9	14
cottonseed	4.2	5.9	41
groundnut	4.2	5.7	35
palmkernel	2.0	3.5	77
coconut	3.0	3.3	9
corn	1.7	2.7	64
olive	2.0	2.1	6
fish	1.2	1.3	11
sesame	0.71	0.94	33
linseed	0.64	0.88	37
castor	0.45	0.55	22
total	90.6	132.1	46

Figures from *INFORM* 1994, **5**, 715 (taken from *Oil World 2012*, 1994).
[a]Based on more accurate figures quoted in *INFORM*.

Table 3.3 World production of palm oil over the period 1965–2005 (millions of tonnes)

Area	1965	'70	'75	'80	'85	'90	'95[a]	00[a]	05[a]
Malaysia	0.16	0.43	1.26	2.58	4.13	6.41	–	–	–
Indonesia	0.16	0.22	0.41	0.69	1.20	2.10	–	–	–
other	1.04	1.08	1.09	1.21	1.41	2.41	–	–	–
Total	1.36	1.73	2.76	4.48	6.74	10.92	13.89	17.50	21.41

[a]Forecast
Source: *Oil World Annual 1991* and *Oil World 1958–2007* (1988), ISTA Mielke, Hamburg. Indonesian production is expected to exceed that of Malaysia from around 2000 onwards.

Table 3.4 Yield of oil and meal from palm, soya, and rape

Source	Tonnes/hectare/annum	
	Oil	Meal
palm:		
fruit	4.2	–
kernel	0.4	0.55
soybean	0.4	1.6
rapeseed	1.0	1.5

encouraged to reduce their total fat intake and to have the appropriate balance of saturated, monounsaturated, and *n*-6 and *n*-3 polyunsaturated acids (section 9.3) there are still heavily populated communities whose main requirement is for calories which are most efficiently provided by fat of any kind. The non-food uses of these materials may also increase in the next few years with increasing production of crops for industrial purposes (section 3.7).

3.2 The major vegetable oils and fats

An account will now be given of the major sources of oils and fats starting · with those of plant origin and presenting them in alphabetical order. Reference has already been made to the production levels of the more important sources and their chemical composition is detailed in a later section.

3.2.1 *Blackcurrant, borage, and evening primrose oils*

Interest in these three minor oils, particularly evening primrose oil, is based on the fact that they all contain γ-linolenic acid (*n*-6 18:3), which is an important essential fatty acid believed to be beneficial in the treatment of multiple sclerosis, arthritis, eczema, premenstrual syndrome, and other diseases. Details of their fatty acid composition are given in Table 3.5.

3.2.2   Castor oil (Ricinus communis)

Castor oil is unique among commercial oils and fats in that it contains a very high level (around 90%) of a single acid with an unusual structure, namely ricinoleic (12-hydroxyoleic) acid. This acid or its salts, before or after hydrogenation, has important lubricating properties and is used to manufacture greases and cosmetics. The oil or the acid is also converted on a commercial scale to a range of valuable materials (section 1.12.1).

3.2.3   Cocoa butter (Theobroma cacao)

This solid fat, obtained from the cocoa bean along with cocoa powder, is in high demand because of its characteristic melting behaviour which makes it suitable as the fat ingredient of chocolate. It has a sharp melting curve just below body temperature and its rapid melting produces a cooling sensation in the mouth. This property is related to the glyceride compositions of the oil. The fat is rich in palmitic (24–30%), stearic (30–36%), and oleic acid (33–39%). The main glycerides are of the type glycerol 1,3-disaturated 2-oleate (POP, POSt, and StOSt; P = palmitic, St = stearic, and O = oleic) and comprise 65–80% of the fat (see section 3.5). A few other natural oils showing similar melting behaviour are rich in these same triacylglycerols although the proportion of palmitic and stearic acids may vary (Table 3.6). Cheaper substitutes of cocoa butter have also been derived from less expensive sources (sections 6.1.4 and 8.3.4).

3.2.4   Coconut oil (Cocos nucifera)

Coconut oil comes from copra from coconut trees grown especially in the Philippines, Indonesia, India, Sri Lanka, and Papua New Guinea. Like

Table 3.5 Component acids of oils containing γ-linolenic acid (typical results, % wt)

Source	16:0	18:0	18:1	18:2	18:3 (n-6)	Other
blackcurrant[a]	7	2	11	47	17	18:3 13[d] 18:4 3[d]
borage[b]	10	4	16	38	23	20:1–24:1 8
evening primrose[c]	7	2	9	72	10	–

[a]Ribes niger, [b]Borago officinalis, [c]Oenothera biennis and O. lamarkiana.
[d]Both are n-3 acids.

Table 3.6 Component acids of cocoa butter and some similar fats (typical results, % wt)

Fat		16:0	18:0	18:1	Other
cocoa butter	*Theobroma cacao*	26	34	35	5
Illipe butter[a]	*Shorea stenoptera*	18	46	35	1
sal fat	*Shorea robusta*	5	43	40	12
shea butter	*Butyrospermum parkii*	3	44	46	7

[a] Also called Borneo tallow.

Table 3.7 Component acids of lauric oils (typical results, % wt)

Source	8:0	10:0	12:0	14:0	16:0	18:0	18:1	18:2	Other
coconut	8	7	48	16	9	2	7	2	1
palmkernel	4	4	45	18	9	3	15	2	0

Source: BNF Task Force Report (see Bibliography).

palmkernel oil it is rich in lauric acid (45–50%) and in other short and medium-chain acids (C_8–C_{14}), and these so-called lauric oils are in high demand with both food and non-food uses (Table 3.7).

3.2.5   *Corn oil* (maize oil, *Zea mays*, Table 3.9)

This oil is extracted from the germ after it has been separated from the grain. The latter is used for fodder and for the production of corn starch, sugar, and syrup. The refined oil has good oxidative stability and is used in salads, for frying and, after partial hydrogenation, in margarines. Growing demand for bio-ethanol as a fuel additive is expected to lead to an increased production of maize and hence of corn oil (see Table 3.2b).

3.2.6   *Cottonseed oil*

The cotton plant (*Gossypium hirsutum*, Table 3.9) is grown for its fibre with oil being produced only as a by-product and representing only 11–12% of the gross value of the crop. The major growing areas include China, USA, FSU, India, and Pakistan. Cottonseed oil is characterized by a relatively high content of palmitic acid (20–30%) and is used as a salad oil, a frying oil, and in spreads.

3.2.7   *Gourmet oils*

These are also called speciality oils or exotic oils. They are produced in low-volume and are available in Europe and North America. They include almond, apricot, avocado, grapeseed, hazelnut, macadamia, mustard, pecan, passion flower, pistachio, pumpkin, sesame, walnut, and wheatgerm, in addition to some of the bulk oils (coconut, corn, cottonseed, olive, groundnut, palmkernel, rape, safflower, soya, and sunflower). They are sold as food oils, often on account of their distinctive flavours, and are also used in toiletries, cosmetics, and pharmaceuticals. Careful sourcing of seeds/nuts is necessary to obtain products of the highest quality and this must be associated with appropriate extraction procedures and (if desired) refining. The plant should be of stainless steel and nitrogen blanketing should be used wherever possible. Oils may be cold-pressed, pressed at higher temperatures, and/or subjected to solvent extraction. Bearing in mind the need for low or zero levels of pesticides, attention may have to be given to growing the oils/nuts under conditions of 'organic farming'.

Typical fatty acid compositions of these oils are given in Table 3.8. The oils may be categorized as being oleic-rich (almond, apricot, avocado, hazelnut, macadamia, pecan, and pistachio), linoleic-rich (grapeseed, passionflower), containing moderate levels of both oleic and linoleic acid (sesame), containing α-linolenic acid in addition to oleic and linoleic (walnut and wheatgerm), and containing monoenes higher than 18:1 (mustard).

3.2.8 Groundnut oil (peanut oil, monkey nut oil, *Arachis hypogaea*, Table 3.9)

This grows in tropical and subtropical climates and is produced particularly in China, India, and USA. Its major acids – palmitic (about 13%), oleic (around 38%), and linoleic (around 41%) – are accompanied by small amounts of saturated and monoene C_{20}–C_{24} acids (up to a total of about 5%). The oil is highly favoured as a salad and frying oil.

Table 3.8 Component acids of some speciality oils (typical results, % wt)

Source	16:0	18:0	18:1	18:2	18:3	Other
almond	7	2	61	30	–	–
apricot	5	1	60	32	–	2
avocado	12	1	71	14	1	1
grapeseed	7	4	16	72	1	–
hazelnut	7	2	69	21	–	1
macadamia	8	2	56	3	–	9, 16:1 22
mustard	3	1	23	9	10	3, 20:1 8, 22:1 43
pecan	6	2	67	22	1	2
passion flower	9	2	13	75	–	1
pistachio	9	2	69	18	–	2
sesame	10	6	40	43	1	–
walnut	8	3	18	58	8	5
wheatgerm	13	2	19	60	5	1

3.2.9   Linseed oil (Linum usitatissimum, Table 3.9)

This is an industrial oil characterized by high levels of α-linolenic acid (45–60%) and used to produce oil-based paints and linoleum. The use of the latter, in decline for several years, is increasing because it has a number of advantages. Manufactured mainly from cork and linseed, it is seen as a 'green' product. It is also hard-wearing, has a long life, is more fire resistant than vinyl floor covering, and presents no problems in disposal. Linseed is grown in Russia, North America, India, and Argentina and is becoming more popular in Europe, especially the UK. (See also linola oil, section 3.7.5).

3.2.10   Olive oil (Olea europea, Table 3.9)

Olive oil is a tree crop grown predominantly in the Mediterranean region. Available in many grades, it is characterized by its high level of oleic acid (65–85%) and by its distinctive flavour. It has high oxidative stability because of the presence of natural antioxidants.

3.2.11   Palm oil and palmkernel oil (Elaeis guinensis, Table 3.9)

The palm produces two oils with quite different fatty acid composition. The soft fleshy mesocarp yields palm oil which is rich in palmitic (about 40%) and oleic acids (about 40%) and also contains useful levels of linoleic acid (around 10%). The seed or kernel produces a second oil at 10–20% of the level of palm oil. Palmkernel and coconut oils are the two major lauric oils with about 50% of lauric acid and other medium-chain acids at lower levels (Table 3.7). The range of uses of palm oil and palmkernel oil is extended by fractionation into a less-soluble, high-melting fraction (the stearin) and more-soluble, lower-melting fraction (the olein). The oil palm grows in the tropical areas of south-eastern Asia, Africa, and America. The tree produces fruit 3–5 years after planting and continues to do so economically for a further 15–20 years. The present yield of oil at about 4–5 tonnes/hectare of palm oil and about 0.5 tonnes of palmkernel oil is greater than that for any commercial oil. Palm oil and its fractions find wide use as edible oils.

3.2.12   Rapeseed oil (Brassica campestris and B. napus, Table 3.9)

The rape now grown and traded for edible purposes is either single-low (low in erucic acid) or double-low (low in erucic acid and in glucosinolates). These varieties have been developed from high-erucic rape by seed breeding. Erucic acid is contra-indicated on health grounds while glucosinolates remaining in the meal act as anti-nutrients. They reduce the level at which rapeseed meal can be mixed with other feed meals. High-erucic rape is still

Table 3.9 Component acids of the major vegetable oils (typical results, % wt)

Source	16:0	18:0	18:1	18:2	18:3	Other
corn	13	3	31	52	1	–
cottonseed	27	2	18	51	trace	2
groundnut	13	3	38	41	trace	$C_{20-24}5$
linseed	6	3	17	14	60	–
olive	10	2	78	7	1	2
palm	44	4	40	10	trace	2
palm olein	40	4	43	11	trace	2
palm stearin	47–69	~5	20–38	4–9	trace	–
rape (high-erucic)	3	1	16	14	10	20:1 6, 22:1 50
rape (low-erucic)	4	2	56	26	10	20:1 2
rice bran	16	2	42	37	1	2
safflower (high-linoleic)	7	3	14	75	–	1
safflower (high-oleic)	6	2	74	16	–	2
sesame	9	6	38	45	1	1
soybean	11	4	22	53	8	2
sunflower (high-linoleic)	6	5	20	69	trace	–
sunflower (high-oleic)	4	5	81	8	trace	2
tall oil	5	3	46	41	3	2

Source: F.B. Padley, *The Lipid Handbook*, 2nd edn (1994), Chapter 3.

grown for non-food purposes (sections 8.7.1 and 10.6) and attempts are being made to produce varieties with even higher levels of this acid. Rape grows in northern climes and is the only vegetable oil cultivated on a large scale in northern Europe (Sweden, Germany, France, Britain, Poland, etc.). It is also grown extensively in China, India, and Canada (where it is known as Canola oil). The rape lends itself to genetic manipulation and genetically engineered rapeseeds rich in lauric acid (a lauric oil) or in stearic acid (a cocoa butter-like fat) are in advanced stages of development. Serious attempts are also being made to introduce less common acids such as petroselinic and ricinoleic acid into rapeseed and to develop seed with still higher levels of oleic acid or erucic acid. Edible rapeseed oil is rich in oleic acid (50–65%) but also contains significant quantities of linoleic (20–30%) and linolenic acid (6–14%) and has a lower level of saturated acids than any other commercially available oil. It is frequently subject to partial hydrogenation before use.

3.2.13 Rice bran oil (Oryza sativa, Table 3.9)

The oil from rice bran is a by-product of the rice crop grown primarily for its grain. Rice is the main cereal crop of Asia and is the basis of the diet of at least half of the world. During milling, rice bran and the germ are removed and this furnishes a useful oil on extraction. The oil is quickly hydrolysed leading to a high level of free acid unless the bran is extracted quickly or is heated and dried while awaiting extraction. Good quality rice bran oil is stable to oxidation by virtue of the tocopherols and ferulic esters which it contains. Palmitic, oleic, and linoleic acid make up over 90% of the oil.

3.2.14   Safflower oil (Carthamus tinctorius, Table 3.9)

Normally this is a linoleic-rich oil (55–80%) but high-oleic varieties have been developed (70–75%).

3.2.15   Sesame oil (Sesamum indicum, Table 3.9)

Sesame is grown particularly in China, India, Sudan, and Mexico and contains saturated (around 14%), oleic (33–54%), and linoleic acids (35–59%). Its high oxidative stability is linked to the presence of sesamol which acts as a natural antioxidant.

3.2.16   Soybean oil (Glycine max, Table 3.9)

Soybeans are grown primarily for the protein-rich meal which remains after the oil has been extracted. This meal is used mainly for animal feed. Soybean meal makes up at least 60% of the world meal market and represents 50–70% of the value of the soybean. The balance lies in the value of the oil. The major producing countries are USA, China, Brazil, and Argentina and there is a large intercontinental trade in whole beans as well as in meal and extracted oil. As indicated in Table 3.9 soybean oil is highly unsaturated being rich in linoleic acid (>50%) and also containing some linolenic acid (4–11%). Much of the oil is consumed only after partial hydrogenation, particularly to reduce the level of linolenic acid. Attempts are being made to develop lines where the oil contains less saturated acids and less linolenic acid. The latter is thought to be responsible for undesirable flavour notes that develop in the oil.

3.2.17   Sunflower oil (Helianthus annus, Table 3.9)

This plant grows in temperate zones and is cultivated extensively in former Soviet Union, the European Union (especially France, Spain, and Italy), Argentina, China, USA, and eastern Europe. The seed oil is low in saturated acids, contains virtually no n-3 acids, and is rich in linoleic acid (up to 70%). It is highly favoured, especially in Europe, for its high level of linoleic acid and is an important component of margarines rich in polyunsaturated acid. A high-oleic form of sunflower oil with around 80% of oleic acid has been produced (HighSun™ or Sunola™). This is probably the richest natural source of oleic glycerides.

3.2.18 Tall oil

This by-product of the wood pulp industry is mainly a mixture of resin acids (about 42%) which are cyclic terpenes and of fatty acids (around 45%) which are mainly oleic and linoleic acids (each 40–45%). The two types of material can be separated by distillation. The resulting fatty acids are the cheapest source of such material. It is to be noted that they are available only as free acids and not as glycerol esters (Table 3.9).

3.3 Fats of animal origin

3.3.1 Butter fat

Cow milk fat, consumed mainly as milk or butter, is an important source of dietary lipids even if milk is consumed increasingly in the West in a reduced fat form and butter is replaced by margarine and by reduced fat spreads which may or may not contain butter. Milk fat has a very complex fatty acid composition with high levels of short and medium-chain acids (Table 3.10) and with many uncommon acids, including those with *trans* unsaturation, present at lower levels. Attempts to extend the usefulness of butter fat by fractionation and to find alternative uses for the fat or its fractions are hampered by the higher cost of the starting material compared with that of vegetable oils. The world production of butter (as fat) is around 6–7 mt (see also section 10.2).

3.3.2 Lard and animal tallows

Lard from pigs and tallows from cattle and sheep are produced by rendering of carcass tissue. In addition to the common acids these animal

Table 3.10 Component acids of milk fats (typical results, % wt)

Source	Saturated								Unsaturated				
	4	6	8	10	12	14	16	18	16:1	18:1	18:2	18:3	Other
cow	3	2	1	3	4	12	26	11	3	28	2	1	4
human	0	0	0	1	5	7	27	10	4	35	7	1	3

Source: BNF Task Force Report (see Bibliography).

fats contain low but significant levels of acids with an odd number of carbon atoms, with branched-chains, and with *trans* unsaturation. The major acids are palmitic and oleic with substantial levels of linoleic acid in lard, hexadecenoic acid in beef tallow, and stearic acid in lard and mutton tallow (Table 3.11). Compared with vegetable oils these animal fats are rich in cholesterol and deficient in natural antioxidants. Despite their relatively saturated nature, therefore, they have to be stabilized against oxidation by the addition of natural or synthetic antioxidants.

3.3.3 Fish oils

These make only a minor contribution (about 2%) to the annual production of oils and fats. They come mainly from fish such as herring, capelin, sand eel, menhaden, sardine, and anchovy. The biggest use of fish oil (roughly 70%) is in margarine production after partial hydrogenation. The remainder is burnt for fuel, used in animal feed, or used for industrial purposes. The protein-rich meal remaining after extraction of the oil is generally the more valuable product and is used for animal feed. Because fish oils are the best source of long-chain *n*-3 polyunsaturated acids such as EPA and DHA, attempts are being made to produce oils with low levels of free acid and of oxidation and to use these, without hydrogenation, as dietary health foods or to incorporate them, along with suitable anti-oxidants, into baking and spreading fats (Table 3.12).

Table 3.11 Component acids of animal fats (typical results, % wt)

Source	14:0	16:0	18:0	16:1	18:1	18:2	Other
lard	2	27	11	4	44	11	1
beef tallow	3	27	7	11	48	2	2
mutton tallow	6	27	32	2	31	2	0

Source: F.B. Padley, *The Lipid Handbook*, 2nd edn (1994), Chapter 3.

Table 3.12 Component acids of fish oils (typical results, % wt)

Source	14:0	16:0	16:1	18:0	18:1	18:2[c]	18:4[d]	20:1	20:5[d]	22:1	22:5[d]	22:6
herring[a]	9	15	7	1	16	1	–	16	3	23	–	3
menhaden	8	22	11	3	21	2	–	2	14	2	–	10
capelin	8	9	16	1	8	1	1	17	9	20	1	3
anchovy	8	20	9	3	15	1	2	3	18	2	1	11
sardine[b]	8	16	9	4	11	1	2	3	17	4	3	13
cod (liver)	4	14	12	3	22	1	1	12	7	11	trace	7

[a]Nova Scotia [b]California [c]*n*-6 [d]*n*-3.
Sources: BNF Task Force Report (see Bibliography). F.B. Padley, *The Lipid Handbook*, 2nd edn (1994), Chapter 3.

3.4 Fatty acid composition

The component acids of the oils which have been described are set out in Tables 3.5–3.12 and will be discussed briefly. Since these are natural products they show some variation depending on the growing conditions (plants) or on dietary and environmental conditions (animals). The results provided are offered as typical values and it seems sufficient to quote these to the nearest whole number.

The two important lauric oils are characterized by their high level of C_8–C_{14} saturated acids (70–80%) with lauric acid (45–50%) predominating. The two oils differ slightly in that palmkernel has lower levels of C_8 and C_{10} acids and a higher level of oleic acid compared with coconut oil (Table 3.7).

Most vegetable oils contain palmitic, oleic, and linoleic acid as major components, although sometimes other acids become significant. Palmitic acid is normally below 20% and frequently below 10% but somewhat higher in palm oil (44%) and in cottonseed oil (27%). The remainder is almost entirely oleic and linoleic acid (>80%). Sometimes oleic acid predominates as in olive (78%), low-erucic rape (56%), and the high-oleic varieties of safflower (74%) and sunflower (81%), Linoleic acid is high in soybean (53%), corn (52%), cottonseed (57%), sesame (45%), and groundnut (41%) as well as in the ordinary forms of safflower (75%) and sunflower (69%). Linolenic acid is usually below 1% except in soybean (8%), rape (10%), and linseed (60%).

Two milk fats are reported in Table 3.10. Cow milk fat is particularly rich in C_4–C_{14} acids (25%) but these only add up to 13% in human milk fat which contains higher levels of oleate and linoleate.

Palmitic and oleic acids predominate in animal body fats (Table 3.11) with significant levels of stearic in lard and mutton, 9-hexadecenoic in beef, and linoleic acid in lard.

Fish oils are characterized by the wide range of acids present and particularly by the highly unsaturated members. Any of the following saturated (14:0 and 16:0), monoenoic (16:1, 18:1, 20:1, and 22:1), and n-3 polyenoic acids (especially 20:5 and 22:6) can be major components. The combined monoene acids are particularly high in herring (62%), capelin (61%), and in cod liver oil (57%). The best sources of EPA and DHA are tuna (31%), sardine (30%), anchovy (29%), and menhaden oil (24%).

3.5 Triacylglycerol composition

Natural oils and fats consist mainly of triacylglycerols accompanied by lower levels of monoacylglycerols and diacylglycerols, free fatty acids, sterols and sterol esters, tocopherols, and other minor components.

The acids (or methyl esters) which can be liberated from the glycerol esters are easily analysed by gas chromatography (section 5.2.3) and most accounts of oils and fats contain extensive data on fatty acid composition. Similar data on the component glycerol esters is harder to find and there are several reasons for this. First of all, the chromatographic procedures (GC and HPLC) for quantitative triacylglycerol analysis are not so simple nor so widely employed as the gas chromatography of methyl esters. Secondly, the methods which have been devised to examine this problem usually give only partial information and it is difficult to correlate different kinds of partial results. For example, lipolysis will give information about the fatty acids at the sn-1/3 and the sn-2 positions, gas chromatography gives a separation of groups by molecular weight (carbon number), and HPLC gives a separation of glycerol esters by their partition number. Unless these procedures are carried out sequentially it is difficult to combine results obtained from them. Thirdly, even quite detailed results may not distinguish between stereoisomeric triacylglycerols and finally when the results are fully detailed they are quite complex with a lot of components, many of them present only at low levels. Such data are difficult to present compactly and comparison of different oils is not easy. It should also be added, when comparing results from different laboratories, that small differences in fatty acid composition of different samples of the same oil have a marked effect on triacylglycerol composition.

In the following account emphasis is given to results obtained recently by the more modern techniques described in section 5.2.4, and older

Table 3.13 Triacylglycerol composition of olive oil (% mol)

TAG group[a,b]	Data source		
	1	2	3
SSM	3.2	2.8	7.3
SMM	31.0	26.9	23.3
MMM	45.8	42.8	44.6
SMD	6.0	4.5	5.7
MMD	14.0	11.1	14.8
MDD	–	1.9	1.1
other	–	10.0	3.2

1. F. Santinelli, P. Damiani, and W.W. Christie (1992) *J. Amer. Oil Chem. Soc.*, **69**, 552 (Ag[+] HPLC and stereospecific analysis).
2. R.V. Flor, L.T. Hecking, and B.D. Martin (1993) *J. Amer. Oil Chem. Soc.*, **70**, 199 (HPLC, median value of 99 samples examined).
3. Specifications supplied by LipidTeknik (Stockholm) for a particular batch of olive oil.
[a]S = saturated, mainly palmitic, M = monoene, almost entirely oleic, D = diene, linoleic.
[b]These symbols do not indicate the position of the acyl groups in the glycerol backbone. For example, SMD represents all possible isomers containing one equivalent of palmitic (or stearic), oleic, and linoleic.

information has been ignored. Some further data are given along with the description of the analytical techniques.

Cocoa butter consists almost entirely of triacylglycerols with one (around 85%) or two double bonds (around 15%). Only three glycerol esters are present at levels above 5% namely POP (18–23%), POSt (36–41%), and StOSt (23–31%). It is the high levels of these three glycerol 1,3-disaturated 2-oleates which gives cocoa butter its characteristic melting behaviour.

Since olive oil is very rich in oleic acid (65–82%) it is to be expected that most of its triacylglycerols will contain two or three oleic chains. The results in Table 3.13 from three different laboratories give values of 43–46 for OOO, 23–31% for OOS, and 11–15% for OOL (S = saturated, O = oleic, L = linoleic), and these three components account for 80–90% of the whole oil. OOO (glycerol trioleate) can only be a single isomer: OOS will be mainly the unsymmetrical isomer, the symmetrical glyceride (OSO) will be only a minor component, and OOL will be a mixture of esters in which either oleic or linoleic is attached to the sn-2 position.

Palm oil (Table 3.14) contains more than 80% of palmitic and oleic acids together and therefore most of its glycerol esters will contain one or both of these acids (PPP, PPO, POP, POO, OPO, and OOO). The figures bear this out (about 70%) with the mixed glycerides – POP (and some PPO) and

Table 3.14 Triacylglycerol composition of palm oil (% mol)

TAG	Data source		
	1	2	3
PPP[a]	4.6	6.5	7.0
POP	} 42.8	27.6	} 31.8
PPO		5.3	
POSt	3.5	5.5	5.6
POO	29.2	23.1	22.5 (including PLSt)
StOO	1.1	3.0	2.4 (including StLSt)
PLP	7.5	7.0	9.2
OOO	2.3	3.4	3.9 (including StLO)
POL	7.5	7.3	8.2
other	1.5	11.3	9.4

1. M.H.J. Bergqvist and P. Kaufmann (1993) *Lipids*, **28**, 667 (reversed phase HPLC).
2. S. Takano and Y. Kondoh (1987) *J. Amer. Oil Chem. Soc.*, **64**, 380 (reversed phase HPLC).
3. G.J. Sassano and B.S.J. Jeffrey (1993) *J. Amer. Oil Chem. Soc.*, **70**, 1111 (gas chromatography).
[a]P = palmitic, St = stearic, O = oleic, L = linoleic. These symbols do not indicate the distribution of acids between the α and β positions.

POO – being predominant. Palm oil may be analysed by gas chromatography in terms of C_{48}–C_{54} triacylglycerols depending on the number of
C_{16} and C_{18} acyl chains.

Soybean, rapeseed, cottonseed, and groundnut oils (Table 3.15) are rich
in linoleic and oleic acids and triacylglycerols containing one or both of
these acids predominate (LLL, LLO, LOL, OOL, OLO, and OOO). This
is most apparent in the soybean and rapeseed oils. The glyceride
composition of cottonseed oil is influenced by the high level of palmitic
acid (around 27%) leading to high levels of LLP (28%) and LOP/OLP
(14%). Groundnut oil contains about 21% of C_{16} to C_{24} saturated acids
which are present mainly as glycerol esters of the type OOS (S = saturated
acid, approximately 16%).

3.6 Interchangeability of oils and fats

The idea of interchangeability developed from about 1950 onwards and
was based on the view that many oils and fats were similar enough in
composition, or could be made so by modification and/or blending, to be
used interchangeably. Such a concept could be driven by the desire to

Table 3.15 Triacylglycerol composition of soybean, rapeseed, cottonseed, and groundnut oils (% wt)

TAG	Soybean[a]	Soybean[b]	Rapeseed[b]	Cottonseed[c]	Groundnut[d]
LnLL[e]	7.8	5.6	0.8	–	–
LLL	22.5	32.3	5.1	19.0	5.8
LLO	18.2	23.4	18.9[f]	12.5	26.1
LLP	14.6	17.0	–	27.5	8.4
LOO	9.5	6.9	26.8	3.1	21.5
LOSt	2.4	–	–	1.3	–
LOP	9.4	11.5	2.0	14.0	13.4
OOO	3.0	0.6	41.4	1.6	4.5
OOP	2.0	0.4	2.1	3.1	5.6[g]
LPP	1.5	0.7	–	7.1	1.9
OPP	0.2	–	0.4	2.2	1.0
other	8.9	1.6	2.5	8.6	1.8

[a]Specification supplied by LipidTeknik (Stockholm) for a particular batch of soybean oil.
[b]M.H.J. Bergqvist and P. Kaufmann (1993) *Lipids*, **28**, 667 (reversed phase HPLC).
[c]J.M. Bland, E.J. Conkerton, and G. Abraham (1991) *J. Amer. Oil Chem. Soc.*, **68**, 840 (HPLC).
[d]J.A. Singleton and M.E. Pattee (1987) *J. Amer. Oil Chem. Soc.*, **64**, 534 (HPLC).
[e]Ln = linolenic, L = linoleic, O = oleic, St = stearic, P = palmitic. These symbols do not indicate the distribution of acyl groups on the glycerol backbone. All isomers are included.
[f]Including LnOO.
[g]Also OOSt 4.4%, OOA 4.2%, OOB 1.4%, (A = arachidic, B = behenic).

produce a product of higher quality or to produce products from cheaper starting materials.

In the intervening years much has been learnt about processing fats by hydrogenation, fractionation, and interesterification but, on the other hand, there are higher demands on the physical and nutritional properties of the products and there is a growing demand for fats with a more rigorously defined composition.

In considering interchangeability the following properties have to be considered:

1. The chemical composition of the oils/fats in terms of component fatty acids and component triacylglycerols since these are closely linked to the other important properties.
2. Their physical properties, especially melting behaviour, content of solids, and crystalline form. These are linked particularly with the glyceride composition of the oils/fats.
3. Their nutritional properties, especially the level of saturated, monoene, *trans*-acids, and polyene acids and of the minor components also present (sterols, antioxidants).

3.7 New oilseed crops

In parts of the developed world there is a surplus of some traditional crops and there is a search for break crops and for new crops – including some which produce useful oils. Until now the agricultural industry has produced material mainly for the food industry but in the future more of its crops will be for the chemical industry. Unless a new vegetable oil has some specific property – like oils containing γ-linolenic acid (section 3.2.1) – it will have to compete with existing supplies of oils available at bulk prices. This exerts several constraints. If the new oil contains novel fatty acids then appropriate uses for this will have to be developed. The following factors are significant.

1. The new crop must be easily cultivated, harvested, processed, and marketed. Beyond this, additional costs will result from the need to have separate and distinct harvesting, processing, and marketing facilities.
2. Crops coming from wild species must be domesticated to make them suitable for large-scale farming. Whatever the source, yields of seeds per hectare, oil in the seed, and the desired fatty acid in the oil must all be optimized by seed-breeding and by developing appropriate agricultural procedures. These include the best time for sowing and harvesting, fertilizer needs, and ways of keeping the plant healthy in the face of disease and pests.

3. Eventually the crop must be available in good and reliable yield, in appropriate volume, and at an acceptable price.
4. The demand for and interest in new crops should come from the oleochemical industry as well as the food industry. Traditionally, the oleochemical industry has used lower-grade and cheaper oils than the food industry. It will pay more in return for specific qualities.
5. The demand must be market-led (or at least market-maintained after the first few years). There is an interest in oils containing high concentrations of a single acid such as lauric, oleic, erucic, etc.
6. While a range of agricultural disciplines must be used to meet these requirements, chemists and technologists must assist in the substitution of existing oils by new (improved) oils and in the development of new uses for both existing and new oils.

New crops or existing crops with changed fatty acid composition can be obtained by the domestication of wild plants (e.g. cuphea), by the modification of existing crops by the traditional methods of the plant breeders (e.g. canola oil, linola oil), or by the application of genetic engineering (e.g. production of a lauric oil from rapeseed).

3.7.1 Lauric oils

Glycerides rich in lauric and other medium-chain fatty acids are used extensively in the food industry and in the production of surface-active compounds (including soaps) – usually as products derived from the C_{12} alcohol. Petrochemical and oleochemical sources compete to meet this need. While the availability of palmkernel oil should rise with the increasing supplies of palm oil which are predicted, the supply and price of coconut oil has sometimes been so variable as to arouse interest in alternative sources of lauric acid. Three approaches have been tried: the domestication of *Cuphea* species rich in medium-chain acids, the production of petroselinic acid-containing oils which can be oxidized to adipic and lauric acids, and the application of gene-splitting techniques to rape so that it produces a lauric oil.

3.7.2 Oleic oils

Oleic acid is an important source material for the oleochemical industry. It is obtained mainly from tallow (up to 50%) and also from tall oil (about 45%) and palm oil (about 40%). These are cheap sources but the oleic acid level is not very high and may have to be raised by fractionation. It is also desirable that the content of linoleic acid should be as low as possible since this acid is readily oxidized and promotes the oxidation of oleic acid. Richer sources of oleic acid include the oils from rape (56%), macadamia

(56% along with 22% of 16:1), almond (61%), olive (78%), as well as the high-oleic varieties of safflower (74%) and sunflower (81%) which have already been developed. Other oils of potential value are available from the Jessenia palm (75–80%) and from *Euphorbia lathyris* (80–85%). Attempts are being made to develop this last crop (caper spurge) and also to produce an oleic-rich rapeseed oil (~80%).

3.7.3   Oils containing petroselinic acid

Seed oils of the Umbelliferae are characterized by the presence of petroselinic acid (6c-18:1), often at high levels. Oil content may be in the range 30–60% with petroselinic acid at 50–90%. Some plants from this family are already cultivated for other purposes (carrot, parsnip, parsley, caraway, etc.) so it should be possible to develop strains optimized for yield of oil and of petroselinic acid. The acid melts at 33°C (cf. oleic acid 12°C). It is less soluble in organic solvents and the two acids can be separated by crystallization. It is oxidized to adipic and lauric acids and may have cosmetic and pharmaceutical applications. Attempts are being made in Europe to develop coriander (80% petroselinic acid) into a useful crop. New high-value uses for this acid are required if petroselinic acid crops are going to be successful.

3.7.4   Oils containing higher monoene acids (C_{20}–C_{24})

Erucic acid, available from high-erucic acid rapeseed oil (HEAR), is used extensively as its amide as an antiblock and slip-promotong agent for polyolefins. It can also be oxidized to the C_{13} dibasic acid required for the production of nylon-13,13. Attempts are being made to raise the level of erucic acid from its present level of about 50% to around 70%. Other potential sources of erucic and other long-chain monoene acids include Crambe oil (*Crambe abyssinicum*) which contains 50–55% of erucic acid and is being grown in North America, honesty seed oil (*Lunaria biennis*) with 22:1 around 43% and 24:1 about 25%, and mustard seed oil (*Brassica alba*, 20:1 roughly 8% and 22:1 about 43%).

Another source of higher monoene compounds is jojoba oil which has already been developed as a marketable but low-volume commodity. It is produced by a drought-tolerant plant which withstands desert heat. This takes 3–5 years to first harvest, 10–17 years to full yield, and has a life span of around 100 years. It is grown in south west USA and Mexico mainly but also in Latin America, Israel, South Africa and Australia. Yields are reported to be 1 tonne of oil/acre.

Jojoba oil is not a typical triacylglycerol but a mixture of wax-esters containing mainly 20:1 and 22:1 acids and alcohols. It is thus a mixture of C_{40}, C_{42}, and C_{44} esters with two isolated double bonds. The oil is a replacement for sperm whale oil which is now prohibited in most countries because the sperm whale is an endangered species. At present it is a high-priced oil used mainly in cosmetics but it has good lubricating properties and could be used extensively for that purpose if available in sufficient quantity at an appropriate price. The oil is fairly pure as extracted, has a light colour and, since the double bonds are well separated, it is resistant to oxidation. The oil can be chemically modified by reaction (hydrogenation, stereomutation, epoxidation, sulphochlorination) of the double bonds in both the acid and alcohol portions of the esters.

Meadowfoam oil (*Limnanthes alba*) is another oil rich in C_{20} and C_{22} acids. It comes from a herbaceous annual grown in the Pacific coastal regions of North America. More than 95% of its component acids have chain lengths of C_{20} or C_{22} and include 5c- 20:1 (63%), 5c- 22:1 (12%), 13c-22:1 (3%), and 5c13c- 22:2 (18%). The double bond is mainly $\Delta 5$ which is an unusual position. The oil remains clear at refrigerator temperature and is fairly stable to oxidation. Potential uses of the oil include conversion to a liquid wax, hydrogenation to a solid ester, use of the isopropyl ester as a cosmetic, use of its glycerol monoester and propylene glycol esters as surface active compounds, vulcanization to a factice, and conversion to amide and dimer acid.

3.7.5   Oils containing C_{18} polyene acids

Reference has already been made to linseed oil which is naturally rich in linolenic acid. Plant breeders in Australia have produced a seed variety which gives an oil high in linoleic acid and with less linolenic acid:

	18:1	18:2	18:3	Other
linseed oil	17	14	60	9
linola	16	72	2	10

This is to be called linola oil and provides a linoleic-rich oil which can be grown in the same northern regions as canola oil (rapeseed). This is being grown in Australia and will be grown as seed becomes available in Canada and in Europe. The new crop has been developed in a little over 10 years and illustrates the fact that it is generally quicker to modify already domesticated crops than to start with wild species.

Another potential new crop is calendula oil (marigold) which contains about 55% of calendic acid (8t10t12c- 18:3). This oil with its conjugated triene acid will be a drying oil like tung oil.

Reference has already been made to three oils containing γ-linolenic acid (6c9c12c- 18:3) (section 3.2.1).

3.7.6 Oils containing epoxy acids

The use of epoxidized soybean oil and linseed oil as stabilizer–plasticizer for PVC is well established. Vernolic acid (12,13-epoxyoleic) occurs at high levels in the seed oils of *Vernonia galamensis* (73–78%), *Euphorbia lagascae* (62%), and *Stokes aster* (65–79%) and attempts are being made to obtain commercial quantities of the first two of these. This acid could provide a source of C_9 and C_{12} dibasic acids, could be fully epoxidized for use with PVC, or be used to produce epoxy coatings.

3.7.7 Oils containing hydroxy acids

Lesquerella fondleri produces oil (25%) which contains ~55% of lesquerolic acid (14-OH 11c-20:1). This is the bishomologue of ricinoleic acid present in castor oil. These two hydroxy acids and the oils from which they are derived should have similar physical properties but will give different products on pyrolysis (tridecenoic acid rather than undecenoic) and on alkali-fusion (the C_{12} dibasic acid in place of the C_{10} member).

Dimorphotheca oil contains dimorphecolic acid (9-OH, 10,12-18:2). This hydroxy conjugated diene acid is easily dehydrated to a mixture of conjugated trienes.

3.8 Phospholipids: sources and composition

All oils and fats in their crude state probably contain phospholipids as minor components. These are largely removed during the degumming stage of refining when they concentrate as a crude mixture of phospholipids and triacylglycerols, generally called lecithin. This is commercially available mainly from soybean but also from rape, corn, and sunflower and is produced at various grades of purity and as concentrates of individual phospholipids. Egg yolk is also a significant source of phospholipids. These are important surface-active compounds extensively used in the food, pharmaceutical, and cosmetic industries.

The major phospholipids present in crude lecithins are usually phosphatidylcholines (PC), phosphatidylethanolamines (PE), and phosphatidylinositides (PI) along with phosphatidic acids (PA) and other minor phospholipids. The proportions of these are somewhat variable. One commercial supplier reports the lecithin from soybean oil to contain PC (20–23%), PE (16–21%), and PI (12–18%) with the balance being mainly triacylglycerols and some minor phospholipids. Values reported for a sunflower lecithin are PC (41%), PE (17%), PI (23%), PA (3%), and others (16%). Higher-quality products containing 90–99.9% of one phospholipid class are also available.

The component fatty acids of these phospholipids vary with their biological source and their lipid class. This is apparent in the figures in Table 3.16.

Recent studies have provided some information about the pairing of the fatty acids in the phospholipids, i.e an analysis of molecular species. As an example the molecular species observed in the *sn*-1 and *sn*-2 positions of PC, PE, and PI in cod roe are detailed in Table 3.17. These three phosphatidyl esters have different compositions: PC and PE are richer in *n*-3 acids (20:5 and 22:6) while PI is richer in *n*-6 acids (20:4).

3.9   Minor components

In addition to glycerol esters most oils and fats contain small quantities of other materials including hydrocarbons such as squalene and the carotenes, sterols, tocopherols and tocotrienols, colouring matters (chlorophyll and carotenes), and vitamins A (retinol), D (cholecalciferol), E (tocopherol), and K (menaquinones). The levels of these components may be reduced during the refining processes.

The triterpene squalene ($C_{30}H_{50}$, Figure 3.1) is present at low levels in most vegetable oils. It attains a higher level in olive oil (0.14–0.70%) and still higher levels in some marine oils (e.g. shark liver oil).

Palm oil is a rich source of carotenoids (0.50–0.70% mainly the α and β isomers at 29–36 and 55–62%, respectively) but much of this is removed during refining. A new process consisting of pretreatment with phosphoric acid and bleaching earth followed by deodorization and deacidification by molecular distillation furnishes a refined oil which is red and still contains about 80% of the original carotenes. In addition to imparting a yellow-red

Table 3.16 Component acids (% wt) of chromatographically purified phosphatidylcholines (PC) and phosphatidylethanolamines (PE) from various sources

Acid	PC				PE	
	Soya	Rape	Sunflower	Egg	Soya	Egg
16:0	15.2	6.4	10.8	38.5	36.6	15.2
18:0	2.8	0.4	3.9	9.2	2.1	28.0
18:1	7.4	55.8	12.1	31.7	9.5	20.9
18:2	67.7	31.0	71.5	11.8	50.2	8.6
18:3	6.4	4.4	0.2	–	1.4	–
20:4 (*n*-6)	–	–	–	1.7	–	9.6
22:6 (*n*-3)	–	–	–	5.2	–	13.2
minor	0.5	2.0	1.5	1.9	0.2	4.5

Source: Specifications supplied by LipidTeknik (Stockholm) a producer of chromatographically purified phospholipids. These figures relate to particular batches of product.

Table 3.17 Molecular species of phospholipids in cod roe

Fatty acids				
sn-1	sn-2	PC[a]	PE[a]	PI[a]
16:0	20:5 (n-3)	15.2	8.5	2.6
18:1	20:5 (n-3)	6.1	12.8	5.1
16:0	22:6 (n-3)	31.0	14.8	5.9
18:1	22:6 (n-3)	14.9	25.7	9.4
16:0	20:4 (n-6)	2.6	3.2	6.2
18:1	20:4 (n-6)	1.9	2.2	16.7
18:0	20:5 (n-3)	0.9	2.5	7.5
20:1	22:6 (n-3)	0.6	5.2	1.6
18:0	20:4 (n-6)	0.7	0.7	36.7
other components		26.1	24.4	8.3

[a]PC = phosphatidylcholines, PE = phosphatidylethanolamines, PI = phosphatidylinositols.
Adapted from M.V. Bell (1989) *Lipids*, **24**, 585.

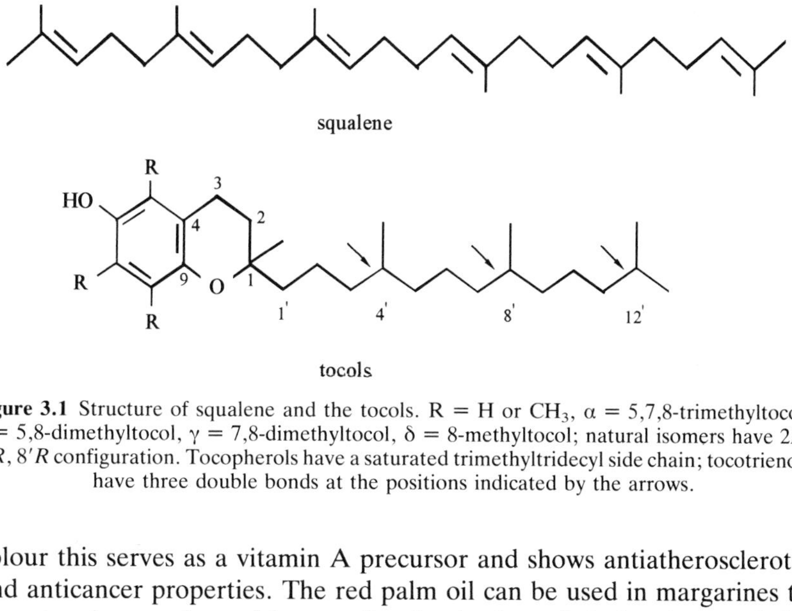

Figure 3.1 Structure of squalene and the tocols. R = H or CH_3, α = 5,7,8-trimethyltocol, β = 5,8-dimethyltocol, γ = 7,8-dimethyltocol, δ = 8-methyltocol; natural isomers have 2R, 4'R, 8'R configuration. Tocopherols have a saturated trimethyltridecyl side chain; tocotrienols have three double bonds at the positions indicated by the arrows.

colour this serves as a vitamin A precursor and shows antiatherosclerotic and anticancer properties. The red palm oil can be used in margarines to impart colour and provide pro-vitamin A, for salad dressing, and for shallow frying.

Among sterols cholesterol is predominant in animal lipids and the following values have been reported: lard 0.37–0.42%, beef fat 0.08–0.14%, mutton tallow 0.23–0.31% and butter fat 0.2–0.4%. Hen eggs contain about 300 mg of cholesterol per egg corresponding to about 5% of total lipid. In the UK the average daily consumption is about 300–400 mg derived mainly from eggs (~150 mg), meat (~100 mg), milk (~40 mg), and spreads (~30 mg) and from foods containing these materials.

Vegetable oils contain little or no cholesterol but instead have brassicasterol (characteristic of rapeseed oil at about 0.06%), campesterol (0.05–0.40%), stigmasterol (0.05–0.30%), β-sitosterol (0.1–0.8%), $\Delta 5$-avenasterol (0.01–0.10%) and $\Delta 7$-stigmasterol (0.01–0.10%). Particularly high values are present in coffee seed oil, corn oil, rice bran oil, and wheatgerm oil. Some values are given in Table 3.18.

Among 4-methylsterols obtusifoliol, gramisterol, and citrostadienol are significant components in vegetable oils and cycloartenol, α- and β-amyrin, butyrospermol, cycloartenol, lupeol, 24-methylenecycloartenol, cyclobranol, and cycloaudenol are triterpene alcohols identified in vegetable oils.

Another important group of materials present in vegetable oils are the tocols (tocopherols and tocotrienols, Figure 3.1). These compounds show two valuable properties: they have vitamin activity (vitamin E) and they are powerful natural antioxidants. These two properties are not identical. For vitamin E activity the order is α (1.0)$>\beta$(0.5)$>\gamma$(0.1)$>\delta$(0.03) and activity is usually expressed in α-tocopherol equivalents. For antioxidants this order is reversed. Some typical values are given in Table 3.19. Among readily available oils palm is a good source of vitamin E because of the high level of the α isomer whereas the soybean tocopherols are effective antioxidants by virtue of the high levels of the γ and δ isomers. The effectiveness of the tocotrienols is less fully understood.

3.10 Waxes

Natural waxes are important compounds with a wide range of uses in the food, pharmaceutical, and cosmetic industries. They are used for the protection of surfaces (shoes, automobiles, furniture, paper, etc.). They come from animal (beeswax, wool wax, sperm whale oil, orange roughy oil) and from vegetable sources (candelilla, carnauba, rice bran, sugar cane, jojoba) and may be solid or liquid. In general the term wax is applied to water-resistant materials made up of hydrocarbons, long-chain acids and

Table 3.18 Sterols in the major vegetable oils

Source	Total sterols (%)	Major sterols (% of total)		
		Campesterol	Stigmasterol	Sitosterol
palm	0.04–0.05	21–24	11–12	57–58
rape	0.48–1.10	18–39	0–1	45–58
soybean	0.18–0.40	16–24	16–19	52–58
sunflower	0.25–0.45	7–13	9–11	56–63

Adapted, from *The Lipid Handbook*, 2nd edn (1994).

Table 3.19 Tocols in the major vegetable oils (mg/kg)

Source	Tocopherols				Tocotrienols		Vitamin E activity[a]
	α	β	γ	δ	α	β	
palm	256	–	316	70	146	3[b]	335
rape	210	1	42	–	–	–	215
soybean	75	15	797	266	2	–	171
sunflower	487	–	51	8	–	–	492
walnut	563	–	595	450	–	–	636
wheatgerm	1330	71	260	271	26	18	1736

– Indicates <1.
[a] As α-tocopherol equivalents.
[b] Also γ-tocotrienol 286.
Adapted, from *The Lipid Handbook*, 2nd edn (1994).

alcohols, the esters derived from these, and ketones. The term wax-ester is confined to the esters of long-chain acids with long-chain alcohols.

Beeswax (from *Apis mellifera*) consists predominantly of wax esters (70–80%), free acids (10–15%), and hydrocarbons (10–20%). These last are mainly C_{25}–C_{33} alkanes (odd numbers only), the free acids are even-chain compounds (mainly C_{28}–C_{34}), and the esters are largely C_{40}–C_{48} compounds based on palmitic acid, oleic acid, and ω2 and ω3-mono-hydroxy palmitic and stearic acids.

Wool wax is recovered from wool during the scouring process. It contains wax esters (48–49%), sterol esters (32–33%), lactones (6–7%), triterpenoid alcohols (4–6%), free acids (3–4%), and free sterols (1%). The acids in the wax ester are straight-chain and branched-chain (*iso* and *anteiso-*) with some hydroxy acids.

Sperm whale oil which is mainly wax esters is now a proscribed product and is replaced by synthetic products or by jojoba oil.

Among vegetable waxes are candelilla (from *Euphorbia antisyphylitica*), carnauba (from *Copernica cerifera*) and the waxes that can be recovered from rice bran or sugar cane. Jojoba oil, which is virtually pure wax ester, has been described in section 3.7.4.

Bibliography

Applewhite, T.H. (1990) *World Conference on Oleochemicals into the 21st Century*, AOCS, Champaign.

Baumann, H. *et al.* (1988) Natural fats and oils – renewable raw materials for the chemical industry, *Agnew Chemie*, **27**, 41–62.

Bourgeois, C. (1992) *Determination of Vitamin E: Tocopherols and Tocotrienols*, Elsevier, London.

British Standard, BS 6406: 1993, ISO 5507: 1992, *Nomenclature for Oilseeds* (names in English, French, and Russian).

Green, A.G. and J.C.P. Dribnenki (1994) Linola – a new premium polyunsaturated oil, *Lipid Technol.*, **6**, 29–33.

Grummer, R.R. (1993) Production and consumption of animal byproducts from the rendering industry, *Lipid Technol.*, **5**, 114–118.

Gunstone, F.D. (1992) Gamma-linolenic acid – occurrence and physical and chemical properties, *Prog. Lipid Res.*, **31**, 145–161.

Gurr, M.I. (1992) *Role of Fats in Food and Nutrition*, 2nd edn, Elsevier, London.

Hase, A. and S. Pajakkala (1994) Tall oil as a fatty acid source, *Lipid Technol.*, **6**, 110–114.

Horrobin, D.F. (1992) Nutritional and medical importance of gamma-linolenic acid, *Prog. Lipid Res.*, **31**, 163–194.

Howard, B. (1993) *Oils and Oilseeds to 1996*, the Economist Intelligence Unit, London.

Jones, L.A. and C.C. King (eds) (1990) *Cottonseed Oil*, National Cottonseed Products Association and the Cotton Foundation, Memphis.

Kamel, B.S. and Y. Yakuda (1994) *Technological Advances in Improved and Alternate Sources of Lipids*, Blackie, Glasgow.

Kolattukudy, P.E. (ed.) (1976) *Biochemistry and Chemistry of Natural Waxes*, Elsevier, Amsterdam.

Mielke, S. (1988) *Oil World, 1958–2007*, ISTA Mielke, Hamburg.

Mielke, T. (1991) *Oil World Annual, 1991*, and later volumes ISTA Mielke, Hamburg.

Moreton, R.S. (ed.) (1988) *Single Cell Oil*, Longman, Harlow.

Murphy, D.J. (ed.) (1994) *Designer Oil Crops – Breeding, Processing, and Biotechnology*, VCH, Weinheim.

Niewiadomski, H. (1990) *Rapeseed – Chemistry and Technology*, Elsevier, Amsterdam.

Padley, F.B. (1994) in *The Lipid Handbook* (eds F.D. Gunstone *et al.*), 2nd edn, Chapman and Hall, London, Chapter 3.

Prentice, B.E. and M.D. Hildebrand (1991) Exciting prospects for flax and linseed oil, *Lipid Technol.*, **3**, 83–89.

Pryde, E.H. *et al.* (eds) (1981) *New Sources of Fats and Oils*, AOCS, Champaign.

Report of the British Nutrition Foundation's Task Force (1992) *Unsaturated Fatty Acids – Nutritional and Physiological Significance*, Chapman and Hall, London.

Robbelen, G. *et al.* (eds) (1989) *Oil Crops of the World – Their Breeding and Utilisation*, McGraw-Hill, New York.

Sayre, B. and R. Saunders (1990) Rice bran and rice bran oil, *Lipid Technol.*, **2**, 72–76.

Schwitzer, M. (1991) Oleochemicals from ricinoleic acid, *Lipid Technol.*, **3**, 117–121.

Shahidi, F. (1991) *Canola and Rapeseed*, Chapman and Hall, London.

Sundram, K. and A. Gapor (1992) Vitamin E from palm oil: its extraction and nutritional properties, *Lipid Technol.*, **4**, 137–141.

Timmermann, F. (1994) Production and properties of medium-chain triglycerides, *Lipid Technol.*, **6**, 61–64.

van Soest, L.J.M. and F. Mulder (1993) Potential new oilseed crops for industrial use, *Lipid Technol.*, **5**, 60–65.

Vietmeyer, N.D. (1985) *Jojoba – New Crop for Arid Lands – New Raw Material for Industry*, National Academy Press, Washington.

4 Processing: Extraction, refining, fractionation, hydrogenation and interesterification

This chapter is devoted to the processes by which raw material from land animals (60–90% of fat in animal tissue), fish (10–20% of whole fish), and vegetable sources (20–70%) yields extracted fat, how this can be refined to a more acceptable food product, and how the materials produced by nature can be further modified to give physical, chemical, and nutritional properties which are more appropriate for the final use of the lipid. Some reference will also be made to handling, transport, and storage at various stages between harvesting and final usage.

4.1 Extraction

Animal fats, from whole animals, depot fats, or viscera, contain active enzymes so the fat should be extracted as quickly as possible. Oils to be extracted from whole fish or from fish waste are subject to the same problems and the highly unsaturated nature of fish oils also make them liable to quick oxidation. Vegetable fruits such as the oil palm have a high moisture content and here also the oil should be extracted as quickly as possible to prevent enzymic hydrolysis. Vegetable seeds are drier and can be safely stored or transported prior to extraction at moisture levels between 7 and 13%.

Fat is recovered from land animals by reduction to suitably sized pieces (hashing) followed by steaming in one of two ways. With wet steam the temperature is raised to about 100°C and the mixture agitated. After settling, the oil layer is first decanted and then centrifuged. With dry steam (indirect steam heating) the product is put through a screw press and the oil/water mixture is separated with a decanter and a centrifuge. Oil is recovered from fish by a similar process of dry rendering. At these temperatures enzymes are deactivated.

Palm fruits are first sterilized at around 130°C to loosen individual fruits from the bunch and to deactivate hydrolytic enzymes. They are then broken up in a digester which produces liquid (oil/water) and wet solid fractions. These are separated in a screw press and the oil is freed from water by centrifuge and by vacuum drying.

Seeds are pretreated in a series of processes which involve cleaning, dehulling or decorticating, size-reduction, cooking at 90–115°C and flaking. The resulting material can then be put through a screw press at 4.5 kg/mm² to squeeze out the oil and leave protein meal still containing 3–6% of oil.

If the seed cake is to be solvent extracted it is sufficient to operate the screw press at ~2 kg/mm². The residual cake is extracted by percolation or immersion in solvent and finally contains only 0.5–1.0% of oil.

When the seeds are not very rich in oil prepressing is unnecessary and the seeds, suitably adjusted in size and moisture content, are extracted directly with solvent. This is generally 'hexane' which is a mixture of hexane and methylpentane. The solvent is recovered for re-use but there is a loss of 8–10 litres/tonne of meal and 2 litres/tonne of oil. The oil content of some common vegetable sources is given in Table 4.1.

It has been suggested that extraction be effected with alcohols or with supercritical carbon dioxide but these are not yet used on a commercial scale.

Attempts are being made to improve extraction by pressing by pretreatment with an enzyme such as *Aspergillus niger* which helps to break down cell walls. Oil extraction then approaches 90% of that achieved by solvent extraction. This system requires less energy, avoids loss of solvent, and produces oil and meal both of higher quality.

4.2 Refining

The conversion of a crude oil to an edible oil involves one or more of the following processes: degumming, neutralization or physical refining,

Table 4.1 Oil content of some plant sources

Source	Oil content
babassu	60–65
copra	65–68
corn	5
cottonseed	18–20
groundnut	45–50
olive	25–30
palm fruit	45–50
palmkernel	45–50
rapeseed	40–45
safflower	30–35
sesame	50–55
soybean	18–20
sunflower	35–45

Quoted with permission: F.D. Gunstone and F.A. Norris (1983) *Lipids in Foods: Chemistry, Biochemistry, and Technology*, Pergamon, Oxford, p. 97

bleaching, and deodorization, which will be described briefly (Table 4.2, Figure 4.1). The free acids can be removed by neutralization with alkali (chemical refining) or by steam refining (physical refining). The tocopherols and tocotrienols are valuable components of natural oils, particularly in those of vegetable origin, where they act as natural antioxidants. The levels

Table 4.2 Refining processes

Process	Procedure	Impurities removed or reduced
degumming	H_3PO_4, H_2O, 70–80°C	phospholipids, trace metals, pigments, carbohydrates, proteins
neutralization	NaOH (other alkali)	fatty acids, phospholipids, pigments, trace metals, sulphur compounds, oil-insolubles, water solubles
washing	H_2O	soap
drying	–	water
bleaching	bentonite etc.	pigments, oxidation products, trace metals, sulphur compounds, traces of soap
filtration	–	spent bleaching earth
deodorization or physical refining	steam under reduced pressure	fatty acids, mono- and diacylglycerols, oxidation products, pigment decomposition products, pesticides
polishing	–	traces of oil insolubles

Adapted from *The Lipid Handbook*, 2nd edn (1994) p.260.

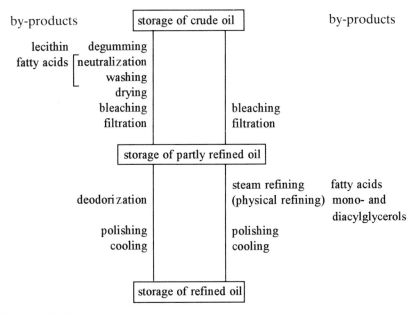

Figure 4.1 Refining processes by chemical (left) and physical procedures (right). Adapted from *The Lipid Handbook*, 2nd edn (1994), p. 260.

of these should be kept as high as possible throughout the refining process. It is also desirable to maintain the nutritional value of the natural oil. This requires keeping stereomutation to the minimum (see section 4.2.4).

4.2.1 Degumming

This involves heating the oil with phosphoric acid when most of the phosphorus-containing compounds are hydrated and give insoluble material which can be removed. This technical grade lecithin (a mixture of phospholipids and triacylglycerols, section 3.8) is a useful commodity in its own right.

4.2.2 Neutralization

This is effected by treatment with alkali and should be carried out with minimum loss of neutral oil. The resulting soap is separated and can be acidified to give fatty acids. Physical refining (steam refining) is an alternative process which removes free acids and mono- and diacylglycerols along with other minor components. This process can be applied only to oils with low levels of phospholipids.

4.2.3 Bleaching

This is carried out mainly to reduce colour and involves heating the oil (80–180°C but mainly 90–120°C), in the absence of oxygen, with a bleaching earth such as bentonitite, fuller's earth, activated carbon or amorphous silica. The level of bleaching agent is between 0.2 and 2.0% of the weight of oil. Palm oil can be heat-bleached at high temperatures in the absence of bleaching earth.

4.2.4 Deodorization or physical refining

These final steps in the refining process are designed to produce an oil with bland flavour and odour and good shelf-life. This requires, particularly, the removal of those oxidation products which are responsible for off-flavour and is achieved by heating at temperatures between 170°C and 250°C under reduced pressure and using a little steam as a sparge. With higher temperatures (especially greater than 220°C) there is some danger of stereomutation. In vegetable oils linolenic acid is particularly prone to this change.

4.2.5 Chromatographic refining (super refining)

Triacylglycerols and wax esters are used in the cosmetics and pharmaceutical industries generally as carriers of more-valuable materials (perfumes, pigments, drugs). It is therefore desirable that they should be free of less-desirable minor components remaining in natural oils even after conventional refining. This can be achieved by chromatography on a multi-kilogram scale and is applied to oils/fats and also to phospholipid preparations.

Most phospholipid samples are mixtures of different phospholipid classes. It is possible to separate these to produce, for instance, products that are entirely phosphatidylcholines or phosphatidylethanolamines. These of course are still mixtures depending on their fatty acid composition and will vary with their natural origin (see section 3.8).

When glycerol esters are purified by chromatography the products have much less colour (about 90% removed), reduced odour and polar impurities, and improved oxidative stability. These factors are important when the oils or fats are to be mixed with pigments or with expensive or delicate fragrances, or when it is desirable that they should be taste-free (use in lipsticks).

4.3 Hydrogenation

In 1897 Sabatier and Senderens demonstrated that olefinic compounds could be reduced with hydrogen in the presence of nickel or some other metallic catalyst. Soon after, Normann applied the process to unsaturated fatty materials. Partial hydrogenation has since developed into a much-used process for modifying liquid oils from oilseed or from fish.

It has been claimed that among all edible fats one third are hydrogenated and only one tenth have been fractionated or interesterified. These proportions may change in the light of perceived concerns about *trans*-acids. The importance of hydrogenation also reflects, in part, the fact that hydrogenation is more appropriate for highly unsaturated oils such as soybean, rapeseed, and cottonseed, and also the fish oils, while fractionation is better applied to palm oil and other more saturated oils. Since partial hydrogenation results in an increase in the melting point of the oil it is referred to as hardening.

The following changes take place when an oil is partially hydrogenated.

1. There is a change in the melting behaviour of the oil. This is a consequence of the increased proportion of saturated and/or *trans*-monoene acids and affects spreadability, oral response, and baking performance.
2. There is an improvement in chemical stability. This results from reduced levels of the polyene acids which are so easily oxidized.
3. There is a reduction in the nutritional value of the product. This is related to the reduced levels of essential fatty acids (α-linolenic and linoleic acids) and questions have also been raised about the influence

of *trans*-monoene acids on coronary heart disease. It is possible to add back essential fatty acids into the final product by blending with appropriate oils.

The chemical changes responsible for these modified properties are detailed in section 7.1.1. It is sufficient to note at this stage that the unsaturated centres may suffer any of three fates. (i) The double bond can react with hydrogen and become saturated: as a consequence of this diene acids are reduced to monoenes and monoene acids become saturated. (ii) The double bond may change configuration and the natural *cis*-isomers become largely *trans*: such acids have a higher melting point than the *cis*-isomers so stereomutation leads to a rise in melting point without any uptake in hydrogen or change of iodine value (measure of average unsaturation). (iii) Interaction between double bond, catalyst, and hydrogen can lead to double bond migration. In addition, linoleic acid, when partially hydrogenated, can give $\Delta 9$ and $\Delta 12$ C_{18} monoenes each of which may react further. It is possible under conditions of an extended selective hydrogenation that the C_{18} monoene esters include the *cis*- and *trans*-isomers from $\Delta 5$ through to $\Delta 15$ (i.e. 22 isomers). In discussions on hydrogenation much is made of the presence of *trans*-isomers which are easily recognized and estimated in several ways: there is rather less discussion of the effect of double bond position partly because there is no simple instrumented way of detecting and measuring this change.

The process of partial hydrogenation is best understood in terms of selectivity, a term which is applied in several ways. First, it is applied in relation to the acyl chains. Considering the C_{18} members it is possible to consider the sequence set out in Figure 4.2 where k_1, k_2, and k_3 represent the rate of reaction for each stage.

Selectivity S_1 is the ratio k_2/k_3 and usually lies between 5 and 100. When S_1 is high then the hydrogenation product will have a high level of monoenes with little change in saturated acyl chains. Selectivity S_2 is the ratio k_1/k_2 and values are generally restricted to 2–3. This measures the greater reactivity of linolenate (triene) over linoleate (diene). There is a particular interest in preferentially saturating the *n*-3 double bond ($\Delta 15$)

$$18:3 \xrightarrow{k_1} 18:2 \xrightarrow{k_2} 18:1 \xrightarrow{k_3} 18:0$$

$$S_2 = \frac{k_1}{k_2} \qquad\qquad S_1 = \frac{k_2}{k_3}$$

Figure 4.2 Selectivity factors (S_1 and S_2) in the hydrogenation of linolenate, linoleate, and oleate.

since unsaturation at this position leads to compounds with undesirable off-flavours. An alternative solution to this problem under active investigation is to breed soybean and rapeseed oils with reduced levels of linolenic acid. Taking these three reaction rates together ratios of around 12.5:7.5:1 (18:3/18:2/18:1) are considered to be non-selective and ratios around 100:50:1 are considered to be selective.

Another aspect of selectivity (S_i) is concerned with the extent of stereomutation (isomerization) compared to saturation, i.e. with the level of *trans*-monoene. A partially hydrogenated fat will frequently contain up to about 35% of *trans*-acids and this can be as high as 50 or even 70% with a sulphur-poisoned catalyst.

Finally, attention has to be given to the selectivity of the process with respect to the glyceride composition of the product. In the simple case of triolein complete hydrogenation will give tristearin but incomplete hydrogenation can give several oleo, elaido, and stearo glycerol esters. These are set out in Table 4.3 along with their melting points. This is S_t selectivity.

Partial hydrogenation is a flexible process and can produce different products depending on (i) the nature of the starting material, (ii) the extent of hydrogenation, and (iii) the selectivity – particularly S_1 and S_i – which influence the proportion of *cis*- and *trans*-monoenes and of saturated acyl chains. These parameters are controlled by the process conditions and the nature of the catalyst.

The progress of hydrogenation can be monitored by any or all of the following: (i) volume of hydrogen consumed; (ii) iodine value (using an

Table 4.3 Structure and melting points of the products of hydrogenation of triolein (O = oleic, S = stearic, E = elaidic)

	O₃	O₂S		OS₂		S₃
structure	⌈ O ├ O ⌊ O	⌈ O ├ O ⌊ S	⌈ O ├ S ⌊ O	⌈ O ├ S ⌊ S	⌈ S ├ O ⌊ S	⌈ S ├ S ⌊ S
mp(°C)	5	23	12	46	42	73
elaido-isomers mp(°C)	41ᵃ	28–30ᵇ 50ᶜ	27ᵇ 50ᶜ	60	-	-

ᵃtrielaido ester (E₃)
ᵇmonoelaido esters (EOS)
ᶜdielaido esters (E₂S)

accelerated process); (iii) refractive index; (iv) solid fat content by NMR measurement; (v) solid fat index by dilatometry; (vi) slip melting point; and (vii) gas chromatography of the methyl esters. Of these the first three are the easiest and quickest to measure.

Hydrogenation is a reaction between gaseous hydrogen, liquid oil, and solid catalyst and takes place at active sites on the surface of the catalyst. For reaction to occur the gas has to be dissolved in the oil, carried to the catalyst, and adsorbed at the active site where it may react with adsorbed oil. The reacted material must then be desorbed and replaced by another molecule of oil. As indicated above and discussed in more detail in section 7.1.1 the adsorbed molecule of fat may undergo reduction, stereomutation, double bond migration, or even be desorbed without change. Important factors are the catalyst (pore diameter, pore length, activity level, amount) the reaction temperature, the pressure of hydrogen, and the degree of agitation (which affects the transfer of hydrogen and oil to and from the catalyst surface). The effect of five important process variables is summarized in Table 4.4. When increased, all five increase the reaction rate. Increasing temperature and catalyst concentration will increase selectivity and *trans* content: increasing gas pressure and agitation will decrease selectivity and *trans* content: finally, increasing catalyst activity will raise selectivity but decrease *trans* content. Put another way: hydrogenation is favoured by a high concentration of hydrogen on the catalyst (increased pressure, increased stirring) while isomerization is favoured by factors leading to increased hydrogen demand which cannot be completely satisfied (increased temperature, more catalyst, a more active catalyst, a more highly unsaturated oil).

The only useful commercial catalyst is nickel, available at a 17–25% level on a support in hardened edible oil or tallow. This preserves the activity of the nickel in a form in which it is safely and easily handled. The catalyst can be recovered and reused but will be less active. Copper catalysts have also been employed. They have a high S_2, promoting hydrogenation of triene esters but they are less reactive than nickel catalysts and copper is a pro-oxidant and must therefore be removed very thoroughly. Reaction with

Table 4.4 Effect (rise or fall) of process parameters on reaction rate, selectivity, and *trans* content during partial hydrogenation

Increase of:	Selectivity	*trans* content	Reaction rate
temperature	↑	↑	↑
catalyst concentration	↑	↑	↑
gas pressure	↓	↓	↑
agitation	↓	↓	↑
catalyst activity	↑	↓	↑

nickel is usually effected at temperatures between 180 and 200°C and at a pressure of about 3 bar. The catalyst is quickly poisoned by fatty acids, soaps, phospholipids, oxidized acids, sulpur compounds, halogen compounds, carbon monoxide, oxygen, and water. As a consequence of this, attention must be directed to the quality of the oil and of the hydrogen used in the hardening process. The oil should be refined and bleached, have a very low soap content and be dry. Hydrogen, produced mainly by electrolysis of water or by hydrocarbon reforming, should be dry and of high purity. In particular the levels of catalyst poisons (sulphur, chlorine, carbon monoxide, oxygen, and water) and of inert gases (nitrogen, argon, methane) should be carefully monitored.

4.4 Fractionation

Fractionation is a procedure for separating oils and fats into two or more components depending on their solubility (in liquid oils or in solvent) and melting point. The less-soluble, higher-melting fractions are called 'stearins' and the more-soluble, lower-melting fractions are called 'oleins'. The products extend the range of use of the original oil or fat. Sometimes both fractions have added value but on other occasions one fraction is of particular value and attempts have to be made to find a use for the less valuable fraction. Fractionation can be repeated to give further separation but this is only practicable when high-value products are obtained such as cocoa butter replacers.

The process is usually carried out in one of three ways:

1. Crystallization from a solvent such as acetone or hexane gives the best separation but is more expensive to operate.
2. In the Lanza or Lipofrac process an aqueous solution detergent is added after crystallization from the melt. The solid stearin, coated by detergent, goes into the aqueous phase and this facilitates separation of the olein and stearin. This is no longer a widely used procedure.
3. The method most commonly employed involves dry fractionation. The completely liquid oil is cooled slowly. This encourages the production of larger and more uniform crystals (mainly in the β or β' form) which are then separated by filtration under reduced pressure (e.g. Florentine filler) or under elevated pressure through a membrane filter.

Some applications of fractionation include the following:

1. Palm oil (mp 21–27°C) is separated into palm stearin (30–35% of original oil, mp 48–50°C) and palm-olein (65–70%, mp 18–20°C). The latter is in high demand as a high-quality highly stable frying oil. The stearin is the less valuable product but it can be used as a hard fat in margarine stock or as an alternative to tallow in the oleochemical

industry as a rich source of palmitic and oleic glycerides. Double fractionation furnishes palm mid-fraction. This is rich in dipalmitoyl 2-oleyl glycerol (POP) which is an important component of cocoa butter replacers.

2. Palm kernel oil gives a valuable stearin which provides a high-quality hard butter or couverture fat.

3. Fractionation of anhydrous milk fat (AMF) gives a more valuable olein and a less-valuable stearin. The olein can be used to produce a soft butter which spreads from the refrigerator or, sometimes after admixture with anhydrous milk fat, to make butter cookies, butter sponges, butter cream, biscuits, ice-cream cones, waffles, and chocolate (for ice-cream bars). The harder fat can be used in puff pastry (section 10.2).

4. Partially hydrogenated soybean oils of varying iodine values can be fractionated to give salad oils, cooking oils, hard butters, and hardstock.

5. Cottonseed is converted to an improved salad oil by winterization in which the more solid glycerides are removed. Originally this was effected merely by storing at winter temperatures. Now it is more controlled and crystallization occurs at 6°C over 48–60 hours.

6. Crystallization at 6–8°C removes a high-melting wax (mp 76–77°C) from sunflower seed oil yielding a product free of haze.

Hydrophilization is a process for obtaining concentrates of oleic acid from fatty acids of tallow or palm oil. After crystallization at about 20°C the crystals are treated with a wetting agent and form an aqueous suspension which can be separated from the liquid fraction by centrifugation. The latter is mainly oleic acid (70–75%) along with palmitoleic, linoleic, and 10% or less of saturated acids.

4.5   Interesterification

The term interesterification covers a wide range of procedures for modifying esters (including triacylglycerols). These are explained and discussed in section 8.3 and this discussion is confined to methods of modifying triacylglycerols either by randomization of the mixture of glycerol esters in a single oil or fat or by a similar reaction involving two or more different oils. Natural oils and fats do not usually have their acyl chains distributed in a random manner (section 3.5) but randomized mixtures are produced by interesterification with a chemical catalyst with consequent modification of the physical properties. More selective interesterification is achieved with enzymic catalysts.

The catalyst normally employed is an alkali metal at a level of 0.1–0.2% or a sodium alcoholate at a 0.2–0.3% level. The true catalyst is believed to be a diacylglyceroxide ion. Since the catalyst is easily destroyed by acid, water, or peroxides the oil(s) to be interesterified must be free of these impurities. Reaction occurs between 25 and 125°C but is usually carried out at 80–90°C over half an hour.

In a modification of this procedure (directed interesterification) the reaction temperature is only 25–35°C. Under these conditions higher melting triacylglycerols crystallize from the reaction mixture and equilibrium is re-established among esters in the liquid reaction mixture.

Interesting developments in this procedure have followed the use of enzymic catalysts usually present on a support such as kieselguhr, hydroxyl apatite, alumina or phenolformaldehyde resin. Reaction in the presence or absence of organic solvent with traces of water at around 40°C takes up to 16 hours. If desired therefore the reaction may be stopped before completion. Some enzymes are non-specific, some are position specific (reaction at positions 1 and 3 only), and some are fatty acid specific (reaction only for acyl groups with a Δ9 double bond). Some of these are listed in Table 4.5.

A few applications of interesterification are described.

1. Lard, with an unusually high level of palmitic acid in the β position, crystallizes naturally in the β form. When randomized the content of 2-palmito glycerol esters is reduced from around 64% to 24% and the interesterified product crystallizes in the β' form with consequent improvement in shortening properties.
2. The crystal structures of margarines based on sunflower oil or on canola oil (rapeseed) along with hydrogenated oil are stabilized in the β' form by interesterification leading to randomization of the glycerol esters.
3. Solid fats containing about 60% of essential fatty acids can be obtained by directed interesterification of sunflower oil and about 5% of a hard fat.

Table 4.5 Enzymes used to promote interesterification

non-specific	*Candida cylindricae*
	Corynobacterium acnes
	Staphylococcus aureus
1,3-specific	*Aspergillus niger*
	Mucor javanicus
	M. miehei
	Rhizopus arrhizus
	R. delemar
	R. niveus
Δ9-specific	*Geotrichum candidum*

4. A margarine made, for example, by interesterification of palm stearin and sunflower oil (1:1) contains no hydrogenated fat and therefore no *trans*-isomers.

5. The use of 1,3-specific enzymes, such as immobilized *Mucor miehei*, is illustrated in the production of high value products used in the formulation of confectionery fats. In one example palm mid-fraction (section 4.4), rich in the glyceride POP, reacts with stearic acid to give an equilibrium mixture of POP, POSt, and StOSt. Exchange is confined to the 1 and 3 positions with no reaction of oleic acid at the 2 positions. Alternatively, triolein (from high oleate sunflower oil) reacts with stearic acid to give StOSt. The products have to be deacidifed and may need to be fractionated to give a final product with the required properties.

Bibliography

Buchold, H. (1993) Enzyme-catalysed degumming of vegetable oils, *Fat Sci. Technol.*, **95**, 300–308.

Carlson, K.F. and J.D. Scott (1991) *Recent developments and trends in processing of oilseeds, fats and oils, INFORM*, **2**, 1034–1060.

Davies, J. (1992) Advances in lipid processing in New Zealand, *Lipid Technol.*, **4**, 6–13.

Deffense, E. (1995) Dry multiple fractionation: Trends in products and applications, *Lipid Technol.*, **7**, 34–38.

Dieckelmann, G. and H.J. Heinz (1988) *The Basics of Industrial Oleochemistry*, Peter Pomp, Essen.

Erickson, D.R. *et al.* (eds) (1980) *Handbook of Soy Oil Processing and Utilization.* American Soybean Association, St Louis, and American Oil Chemists' Society, Champaign.

Farrer, K.T.H. (1990) *The Shipment of Edible Oils*, IBC Technical Services, Byfleet (UK).

Gunstone, F.D. *et al.* (eds) (1994) *The Lipid Handbook*, 2nd edn, Chapman and Hall, London, Chapter 5, pp. 249–318.

Hoffmann, G. (1989) *The Chemistry and Technology of Edible Oils and Fats and their High Fat Products*, Academic Press, London.

Johnson, R.W. and E. Fritz (eds) (1988) *Fatty Acids in Industry*, Dekker, New York.

Kamel, B.S. and Y. Yakuda (1994) *Technological Advances in Improved and Alternate Sources of Lipids*, Blackie, Glasgow.

Lusas, E.W. *et al.* (1991) Isopropylalcohol to be tested as solvent, *INFORM*, **2**, 970–976.

Murphy, D.J. (ed.) (1994) *Designer Oil Crops – Breeding, Processing, and Biotechnology*, VCH, Weinheim.

Paterson, H.B.W. (1985) *Handling and Storage of Oilseeds, Oils, Fats, and Meal*, Elsevier Applied Science, London.

Patterson, H.B.W. (1994) *Hydrogenation of Fats and Oils: Theory and Practice*, AOCS Press, Champaign, USA.

Robbelen, G. *et al.* (eds) (1989) *Oil Crops of the World – Their Breeding and Utilisation*, McGraw-Hill, New York.

Shukla, V.K.S. and F.D. Gunstone (eds) (1992) *Oils and Fats in the Nineties*, Chapters by Tirtiaux (Fractionation), Willner (Fractionation), Coupland *et al.* (Chromatographic purification), Lips (Hydrogenation), and McCrae (Enzymic interesterification), IFSC, Lystrup, Denmark.

Sjoberg, P. (1991) Deodorization technology, *Lipid Technol.*, **3**, 52–57.

Smith, D.D. *et al.* (1993) Enzymatic hydrolysis pretreatment for mechanical expelling of soybean, *J. Amer. Oil Chem. Soc.*, **70**, 885–890.

Sosulski, K. and F.W. Sosulski (1993) Enzyme-aided *vs* two-stage processing of canola, *J. Amer. Oil Chem. Soc.*, **70**, 825–829.

Tirtiaux, A. (1989) Dry fractionation – a proven technology, *Lipid Technol.*, **1**, 17–20.

Willner, T. and K. Weber (1994) High-pressure dry fractionation for confectionery fat production, *Lipid Technol.*, **6**, 57–60.

Wolff, R.L. (1992, 1993, 1994) *trans*-Polyunsaturated acids in refined oils, *J. Amer. Oil Chem. Soc.* **69**, 106–110; **70**, 219–224, 425–430; **71**, 1129–1134.

5 Analytical procedures

5.1 Standard procedures

Tests for describing various properties of a sample of oil or fat are part of
the assessment of its commercial value or authenticity. They are concerned
with the 'quality' of the oil rather than with its 'chemical nature' though
some tests provide pointers to composition. For example, iodine value is
an indication of the average unsaturation of the acyl chains. In some cases
the chemical reactions underlying the tests are not fully understood and
empirical values are obtained. The results are only reliable and of value for
comparison purposes, when the tests are carried out carefully by
standardized methods which have been worked out and approved by
groups of analysts. There are several such organizations providing similar
but not identical directions. These include:

The Association of Official Analytical Chemists	AOAC
The American Oil Chemists' Society	AOCS
The British Standards Institution	BSI
The International Organisation for Standardisation	ISO
The International Union of Pure and Applied Chemistry	IUPAC

and some other national oil/fat organizations.

Full details of how to carry out these tests are not given here. They are
available in publications of the above organizations and in some other
works. The nature and the purpose of the test will be outlined.

5.1.1 Sampling and oil content

The tests to be described later may be carried out on the whole sample or
on oil or fat extracted from the sample. The sample must first be obtained
and it is important to ensure that the sample taken is representative of the
bulk material. If this is not done properly then the effort subsequently
expended on measuring properties is wasted since the real objective is to
learn about the bulk material and not about the sample.

There are several ways of quantitatively extracting lipids from a sample.
The oil content of seeds can be measured by soxhlet extraction usually with
petroleum ether (bp 40–60°C). This provides a sample for further
investigation. There is also a non-destructive spectroscopic method of
measurement based on wide-line NMR (section 6.5.1). More complex

methods are necessary to extract lipid quantitatively from biological samples or from foods. These materials may contain a wide range of lipid types (triacylglycerols, phospholipids, sphingolipids, and glycolipids) along with water, proteins and amino acids, and carbohydrates.

Important methods of extracting lipids from wet tissue are associated with Folch, Lees, and Stanley (1957) and with Bligh and Dyer (1959) and both make use of mixtures of chloroform and methanol as extracting solvent.

In the Folch method ground or homogenized tissue is shaken with a 2:1 mixture by volume of chloroform and methanol and the organic extract is then shaken with an aqueous solution of potassium chloride (0.88%, ¼ volume of organic extract). The upper (aqueous) and lower (organic) layers contain chloroform:methanol:water in the volume ratios of 3:48:47 and 86:14:1, respectively. It is important that the combined phases have an 8:4:3 volume ratio.

The Bligh and Dyer method was developed to extract lipids from fish muscle or other wet tissue which, it is assumed, contains about 88 g of water per 100 g of tissue. The tissue (100 g) is homogenized with chloroform (100 ml) and methanol (200 ml) and, after filtering, residual tissue is homogenized a second time with chloroform (100 ml). The two organic extracts are combined and shaken with aqueous potassium chloride (0.88%, 100 ml). After settling the lipid is present in the lower layer.

It is important that the extraction be carried out on tissue which is as fresh as possible or has been suitably and carefully stored. Also, all lipid fractions must be handled so as to minimize oxidation.

5.1.2 Melting behaviour

Fats are not pure organic compounds with a sharp melting point but mixtures so that at any given temperature the sample may be wholly solid, wholly liquid, or frequently a mixture of solid and liquid. The proportion of these two phases, the change of this proportion with temperature, and the temperature at which no solid remains are important properties related to the use of the material as spreading or confectionery fats. For example, the relatively sharp melting behaviour of cocoa butter at around mouth temperature is an important factor in the use of this material in chocolate. High-melting glycerides give an unpleasant mouth feel. In contrast, it is undesirable to have frying oils or salad oils which deposit solid during storage whether at ambient temperature or on refrigeration. There are several ways of examining phenomena associated with melting.

The 'titre' is the value (°C) which denotes the solidification point of the fatty acids derived from a fat, while the 'slip melting point' is the temperature at which a column of fat (10 ± 2 mm) contained in a capillary tube and immersed under water to a depth of 30 mm, starts to rise.

Different methods vary in the pretreatment of the fat. Of greater value is the determination of the 'solid fat index' by dilatation or the 'solid fat content' by low-resolution ^1H pulse NMR (section 6.5.1). This latter method has now largely replaced dilatometry. The time taken to make the desired measurement by pulse NMR is very short though this may have to be preceded by time-consuming tempering regimes. Measurements made at a range of temperatures give a plot of percentage solids against temperature. The slope of this curve and the temperature at which there is no solid phase provide useful information about the melting and rheological behaviour of the sample under investigation.

5.1.3 Unsaturation

Average unsaturation is measured by reaction with a suitable halogen for a fixed period of time (usually 30–60 min). The excess reagent, after addition of potassium iodide, is titrated with thiosulphate. The reagent most commonly used is iodine monochloride in acetic acid (Wijs' reagent) with the sample dissolved in di-, tri-, or tetrachloromethane. Attempts are being made to develop procedures in which these undesirable chlorine-containing solvents are replaced by cyclohexane. The reaction is complete in a few minutes in the presence of mercuric salts. The result obtained by this accelerated method may be less accurate but there are occasions when a value is required quickly, for example during partial hydrogenation. The iodine value is less informative than the fatty acid profile obtained by gas chromatography and it has even been suggested that iodine values should be calculated from fatty acid profiles. The iodine values of methyl stearate, oleate, linoleate, and linolenate are 0, 85.6, 173.2, and 260.3, respectively.

Sometimes there is a need to know the proportion of *trans*-isomers among the unsaturated components which normally have *cis* configuration. This can be done by infrared spectroscopy (section 6.3) or, if enough care is taken, by gas chromatography.

5.1.4 Acidity, saponification, unsaponifiable, hydroxyl value

The level of free acid in an oil, both before refinement and after, is an important specification for most fat samples. It is measured by titration with standard sodium hydroxide solution.

The amount of alkali required to hydrolyse (saponify) a fat is a measure of the average chain length of the acyl chains and may be reported as 'saponification value' or 'saponification equivalent'. These are related by the expression SE = 56100/SV.

When a natural fat is hydrolysed it gives fatty acids soluble in aqueous alkali, glycerol soluble in water, and other material insoluble in aqueous alkali which can be extracted with an appropriate organic solvent. This material, described as the 'unsaponifiable fraction' generally includes sterols, terpene alcohols, long-chain alcohols, hydrocarbons, etc. (section 3.9). Normally this will be less than 2% of the oil or fat but sometimes it is higher. Wax esters, for example, are hydrolysed to long-chain acids and alcohols and the latter will be part of the unsaponifiable fraction.

There are also procedures for measuring free hydroxyl groups which may be present in mono and diacylglycerols, in free alcohols, sterols, and in hydroxy acids.

5.1.5   Oxidative deterioration and oxidative stability

In common with all other olefinic compounds oils and fats react with oxygen. The whole process is complex (section 7.2) but it is sufficient to recognize at this point that the primary products of oxidation – even when produced by different mechanisms – are olefinic hydroperoxides which break down to a variety of compounds of lower molecular weight (secondary oxidation products). These are mainly aldehydes largely responsible for the development of the undesirable taste and flavour which characterizes rancid fat. At higher temperatures these products are degraded further to short-chain acids (tertiary oxidation products). These changes are summarized in Figure 5.1

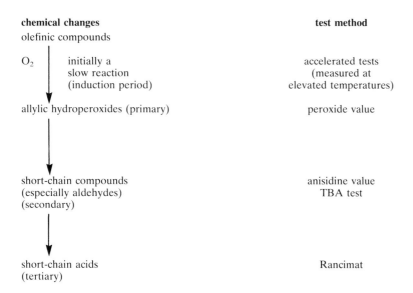

chemical changes   **test method**
olefinic compounds

O_2 initially a   accelerated tests
         slow reaction   (measured at
         (induction period)   elevated temperatures)

allylic hydroperoxides (primary)   peroxide value

short-chain compounds   anisidine value
(especially aldehydes)   TBA test
(secondary)

short-chain acids   Rancimat
(tertiary)

Figure 5.1 Lipid oxidation: chemical changes and test methods.

The most important mechanism of oxidation is autoxidation. This is characterized by an induction period during which oxidation is relatively slow. At the end of the induction period oxidation is quicker and the oil/fat then deteriorates quickly. The length of this induction period is important in estimating the shelf-life of oils and fats and of fatty foods. Because the induction period may be very lengthy at ambient temperature, accelerated tests have been developed in which the oil is held at elevated temperatures and the changes monitored.

There are methods for measuring all these parameters: the primary, secondary, and tertiary products of oxidation and the induction period (Figure 5.1).

The most commonly used method of assessing oxidative status is the (hydro)peroxide value. This is based on the fact that hydroperoxides react with potassium iodide to liberate iodine which can be measured by its reaction with thiosulphate, or electrochemically. The value represents mmol of oxygen per 2 kg of fat. Freshly refined material should have a peroxide value below 1. A fat is rancid at a peroxide level of about 10.

Aldehydes produced as secondary products of oxidation can be measured through determination of the anisidine value or by the thiobarbituric acid test. The first of these is based on assessment of the chromophore at 350 nm produced through reaction of 4-methoxyaniline (anisidine) with aldehydes (principally 2-enals), and the second on the chromophore produced by reaction of malondialdehyde ($OHCCH_2CHO$) or other carbonyl compounds with thiobarbituric acid.

The Rancimat apparatus and the Omnium Oxidative Stability Instrument provide a way of measuring short-chain acids (C_1–C_3) produced during oxidation at temperatures up to 100–120°C.

There are several procedures for measuring the time taken for a fat held at an elevated temperature (up to 100°C) to reach a peroxide value that indicates rancidity. Results of accelerated tests (Schaal, Active oxygen) are assumed to give some indication of the induction period and hence of shelf-life under normal storage conditions. Results obtained at these higher temperatures must always be interpreted with care since it is known that mechanisms of the oxidation reactions change with temperature.

5.2 Lipid analysis

5.2.1 The nature of the problem

The composition of lipid present in an oil or fat, a food, or a biological sample is an important parameter measured regularly in many laboratories. Before discussing the methods available to study this problem it is necessary to have a clear idea of its nature. Lipid composition can be examined at various levels such as:

(a) *What lipid classes are present?* An oil will be mainly triacylglycerols but information may be required on other materials present which include, monoacylglycerols, diacylglycerols, phospholipids, and components of the unsaponifiable material. Phospholipids are mixtures of choline (PC), ethanolamine (PE), inositol (PI), and serine (PS) derivatives. Sphingolipids and glycolipids also comprise several classes (sections 2.3, 2.4, and 2.7).

(b) *What fatty acids (or alcohols or amines) are present?* This is probably the most common question to be addressed and samples of oils and fats often come supplied with a fatty acid profile. Information of this kind will be required to meet labelling regulations which are to be introduced and which many suppliers already provide.

(c) *How are these fatty acids (or other components) combined in the lipids in which they occur?* It is in dealing with this question that matters become complicated. For example, if a wax ester has as few as five component acids and five component alcohols these can be combined to give no less than 25 (5 × 5) different molecular species of wax esters. Can each of these be recognized and measured? The problem is similar with individual phospholipid classes each of which has two acyl chains and with sphingolipids with one acyl chain and one long-chain amine.

The problem is greater with triacylglycerols (section 2.2.3) where the number of potential components can be quite large. Ten different acids could produce 1000 different triacylglycerols (n^3 for n acids). The number of major component acids may be as few as four or five (giving 64 or 125 triacylglycerols) but minor components will raise these numbers. Fish oils contain at least 20 easily identified acids and potentially 8000 triacylglycerols! The number of component acids probably depends as much on the depth of the investigation as on the origin of the sample. This large number has two consequences. First there is an analytical problem. Despite the advances made in recent years it is still not possible to identify and quantitate so many individual species. A difficulty arises from the fact that although several valuable analytical procedures have been described, unless these are carried out sequentially it is not always possible to combine the data obtained in various ways. Finally there is a conceptual problem. Even when 100 or 1000 components have been identified in a lipid it will be difficult to grasp the significance of this information or to compare one lipid with another. We would probably group the components or focus on a few individual members, either because they were very large or because they had some specific property of importance.

The methods employed to examine this problem are mainly chromatographic but also, to some extent, spectroscopic. The latter are discussed in sections 6.2–6.7. The former are summarized in Figure 5.2. Chromatographic methods are essentially adsorption (TLC, HPLC) or partition processes (GLC, HPLC) and the basic procedures can be modified in the ways listed in Figure 5.2. This topic will not be developed on the basis of techniques available but rather on the basis of the problems to be solved.

5.2.2   Lipid classes

The different classes of lipids present in a sample can be identified qualitatively by thin layer chromatography using silicic acid as adsorbent and an appropriate solvent mixture. The method is simple but can only be made quantitative with considerable effort. Some success has been achieved with chromatorods (Iatroscan), or by examining each fraction by gas chromatography after addition of an internal standard and methano-lysis. There are various methods of making separated components visible. Sulphuric acid (conc.) or phosphomolybdic acid can be used for this purpose or these destructive sprays can be replaced by non-destructive materials which are normally required for preparative thin layer chroma-tography. Ethanolic solutions of 2',7'-dichlorofluorescein are widely used for this purpose. Some useful Rf values for various solvent systems are given in Tables 5.1 and 5.2 and some typical chromatograms in Figure 5.3.

Mono-, di- and triacylglycerols can be separated by gas chromatography. For this purpose free hydroxyl groups are usually converted to acetates, trifluoroacetates, or trimethylsilyl ethers. These derivatives are more volatile and less polar than underivatized hydroxy compounds and so behave better on gas chromatography.

High-performance liquid chromatography (HPLC) is an effective form of adsorption chromatography depending on the use of a very uniform, finely divided, micro-spherical (5–10 μm diameter) support of controlled porosity and degree of hydration. This is packed into a stainless steel column (10–30 cm long, 2–4 mm internal diameter) and the sample is eluted under pressure with a single solvent, with a solvent mixture of fixed proportion, or by gradient elution in which solvent composition is changed continuously or in a stepwise fashion. Detection is achieved with an ultraviolet, refractive index, moving band or mass detector. The column may be used in its normal adsorbent mode (polar column and non-polar eluting solvent) or in a reversed-phase manner with, for example, octadecyl silica (ODS). This non-polar column is eluted with polar solvents. These systems separate the various neutral or polar lipids (Figures 5.4 and 5.5).

5.2.3   Component acids (alcohols etc.)

Determination of the component acids of a triacylglycerol or phospholipid sample is probably the most common analytical chromatographic procedure in lipid studies and is a routine operation in most lipid laboratories. The glycerol ester or phospholipid is generally converted to methyl esters (see

adsorption	thin layer chromatography (TLC)
	high performance liquid chromatography (HPLC)
partition	gas liquid chromatography (GLC)
modifications	direct or reversed-phase
	involvement of silver ions
	chiral derivatives or chiral columns

Figure 5.2 Chromatographic systems.

Table 5.1 Rf values of neutral lipids on TLC in three solvent systems

Lipid class	A	B	C
alkanes and alkenes	95	98	98
squalene	95	98	98
sterol esters	90	94	–
wax esters	90	88	–
fatty acid methyl esters	65	77	90
triacylglycerols	35	60	90
fatty acids	18	39	65
long-chain alcohols	15	30	40
sterols	10	19	30
1,3-diacylglycerols	8	21	–
1,2-diacylglycerols	8	15	32
1-monoacylglycerols	0	2	7

Solvent systems: petroleum ether (bp 60–70°C), diethyl ether, and acetic acid: A 90:10:1, B 80:20:1, C 70:30:1.
Adapted from *The Lipid Handbook*, 2nd edn (1994) p. 234.

Table 5.2 Rf values of polar lipids on TLC in four solvent systems

Lipid	A	B	C	D
LPC	–	–	8	–
PI	23	14	11	13
SM	16	–	22	–
PG	48	30	37	39
PE	62	35	41	52
PS	15	6	5	–
Cereb	70–76	–	45–51	58
MGDG	77	51	84	84

Solvent systems:

A	$CHCl_3$, MeOH, H_2O	65:25:4
B	Bu^i_2CO, AcOH, H_2O	40:25:4
C	$CHCl_3$, MeOH, NH_3 (28%)	65:25:5
D	$CHCl_3$, $COMe_2$, MeOH, AcOH, H_2O	10:4:2:2:1

Abbreviations LPC = lysophosphatidylcholine, PI = phosphatidylinositol, SM = sphingo-myelin, PG = phosphatidylglycerol, PE = phosphatidylethanolamine, PS = phosphatidyl-serine, Cereb = cerebroside, MGDG = monogalactosyldiacylglycerol.
Adapted from *The Lipid Handbook*, 2nd edn (1994) p. 235.

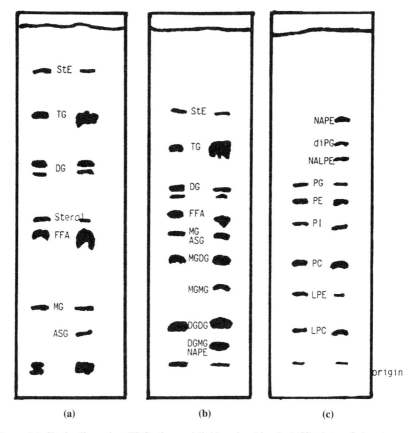

(a)   (b)   (c)

Figure 5.3 Single dimension TLC of cereal lipids using kieselgel 'H' plates. Solvents: a = diethyl ether/toluene/ethanol/formic acid (40:50:1:0.5) for approximately 12 cm, diethyl ether/hexane (8:92) for approximately 18 cm, b = diethyl ether/hexane/formic acid (30:70:1) to 15 cm, chloroform/acetone/formic acid/water (10:90:2:2) 7 cm, diethyl ether/formic acid (99:1) to 12 cm. c = chloroform/methanol/water/tetrahydrofuran/ammonia (65:30:5:3:0.25) to 15 cm. Plates a/b visualized by charring with sulphuric acid; plate c sprayed with Zinzadze reagent to reveal phospholipids only. PC = phosphatidylcholine, PE = phosphatidylethanolamine, DGDG = digalactosyldiacylglycerol, MGDG = monogalactosyldiacylglycerol, PI = phosphatidylinositol, TG = triacylglycerol, PG = phosphatidylglycerol, NAPE = N-acylphosphatidylethanolamine, ASG = acylsterol glucoside, NALPE = N-acylyso-phosphatidylethanolamine, DGMG = digalactosylmonoacylglycerol, LPE = lyso-phosphatidylethanolamine, MGMG = monogalactosylmonoacylglycerol, LPC = lyso-phosphatidylcholine, St = sterol, StE = sterol esters, DG = diacylglycerol, MG = monoacylglycerol, FFA = free fatty acid. Lane 1 on each plate represents standard materials. Reproduced with permission (Hammond, 1993).

section 8.3 for a discussion of suitable procedures) and then submitted to gas chromatography. Some advantages have been claimed for isopropyl esters, such as a better separation of oleate and petroselinate, and butyl and decyl esters have been used for lipids, such as milk fats, containing short-chain acids because of the volatility of the methyl esters. Such

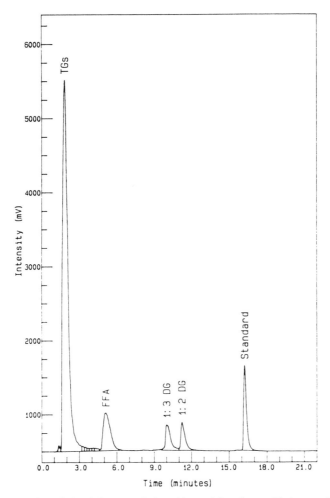

Figure 5.4 HPLC analysis of the neutral glycerides and free fatty acids in a crude palm oil. Monoglyceride is not present in this sample but has an elution time of 21.5 min. The standard is ricinoleyl alcohol prepared from pure ricinoleic acid isolated from castor oil. The column (15 cm × 4 mm i.d. stainless steel) is packed with Merck Lichrosorb S160, 5 μm particle size. Solvent: gradient elution with toluene, hexane, and ethyl acetate containing 5% formic acid. Reproduced with permission (Hammond, 1993).

analyses were first routinely achieved with a packed column (about 150 cm long and 2–4 mm id) using a polar stationary phase and a flame ionization detector linked to a recorder. This gave useful analytical results when due attention was paid to the experimental procedures employed. The packed column has now been largely replaced by capillary columns usually 30–50 m long and 0.2–0.5 mm id. These give better separations than the packed columms. Separation can be effected isothermally (at constant

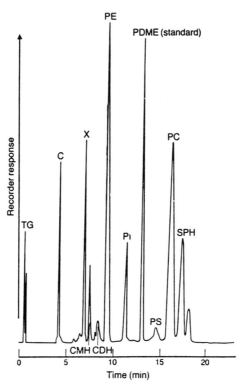

Figure 5.5 HPLC separation of a phospholipid fraction from cows' milk. HPLC separation on a column of silica gel, eluted with a ternary-gradient system and with mass detection. TG = triacylglycerol, C = cholesterol, X = unknown lipid, PE = phosphatidylethanolamine, PI = phosphatidylinositol, PDME = phosphatidyldimethylethanolamine, PS = phosphatidylserine, PC = phosphatidylcholine, SPH = sphingomyelin. Reproduced with permission (Christie, 1987).

temperature) or with a temperature programme. The latter is particularly useful for acid mixtures with a wide range of chain length such as milk fats, lauric oils, and fish oils. Mixtures of *cis*- and *trans*-isomers can also be separated with appropriate columns. The gas phase is usually nitrogen or helium for packed columns and helium or hydrogen for capillary columns.

The identification of chromatographic peaks is based on comparison with authentic samples or with data from the analyses of well-studied oils. For example it is easy to recognize methyl oleate in olive esters, methyl linoleate in sunflower oil, methyl α-linolenate in linseed oil, and the methyl esters of EPA and DHA in the esters from cod liver or other fish oil. Minor components can sometimes be identified by chromatographic behaviour – preferably on two columns of differing polarity – or by combining the separating power of gas chromatography with the identifying power of mass spectrometry in a GC-MS system (see mass spectrometry, section 6.7).

The concept of equivalent chain length (ECL) based on retention behaviour under 'isothermal conditions' has also proved useful. The logarithm of retention time (Rt) of the homologous alkanoates lie on a straight line, the slope of which depends on the polarity of the column. The log Rt of an unsaturated ester placed on this line will give on the appropriate axis a number which is described as the ECL. For example, methyl oleate is eluted after stearate with an ECL in the region 18.2–18.5 depending on the polarity of the column. These values remain roughly constant over the lifetime of the column and can be used in other laboratories for similar columns. Many values have been recorded and tabulated (see Table 1.10) and can be used as an indicator of fatty acid structure. This method is satisfactory for esters which have previously been examined and for which retention data exist.

An example of gas chromatography for the fatty acid esters of pig testis lipids is given in Figure 5.6.

Lipid components other than acids can be analysed in a similar way after they have been converted to appropriate derivatives designed to increase volatility and decrease polarity in order to improve chromatographic behaviour. The most common examples are listed in Table 5.3.

Another procedure for identifying methyl esters is to separate them into fractions depending on the number of unsaturated centres they contain by some form of silver ion chromatography and to examine each fraction subsequently by gas chromatography. In this way components that overlap in the gas chromatography system being used or minor components of uncertain identity may be separated and become significant components of one fraction and be more easily identified.

In one such procedure (Christie, 1987, 1989) a short Bond-elut column is charged with $AgNO_3$-CH_3CN-H_2O and then a range of solvents is used to elute esters with 0–6 double bonds thus: 0 (D), 1 (D, A 90:10), 2 (A), 3 (A, N 97:3), 4 (A, N 94:6) 5 (A, N 88:12), and 6 (A, N 60:40) where D is dichloromethane, A is acetone, and N is acetonitrile.

5.2.4   Triacylglycerols

Triacylglycerol composition can be examined by gas chromatography, by high-performance liquid chromatography, and by the use of silver ion systems.

(a) *Gas chromatography.*   The separation of triacylglycerols of molecular weight around 900 by gas chromatography requires higher temperatures than those used with methyl esters. This was a severe restriction to early studies since the only stationary phases stable at the required temperatures

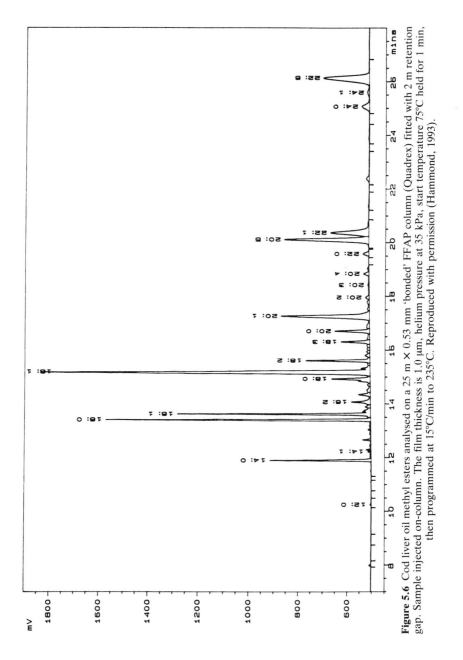

Figure 5.6 Cod liver oil methyl esters analysed on a 25 m × 0.53 mm 'bonded' FFAP column (Quadrex) fitted with 2 m retention gap. Sample injected on-column. The film thickness is 1.0 μm, helium pressure at 35 kPa, start temperature 75°C held for 1 min, then programmed at 15°C/min to 235°C. Reproduced with permission (Hammond, 1993).

Table 5.3 Derivatives used for the gas chromatography of lipid components

Component	Derivative
acids	methyl esters (isopropyl esters)
alcohols	acetates, trifluoroacetates, trimethylsilyl ethers
sugars	poly (trimethylsilyl) ethers
aldehydes	as such or after reduction to alcohols or oxidation to acids
amines	trimethylsilyl derivatives or convert to aldehydes by periodate oxidation

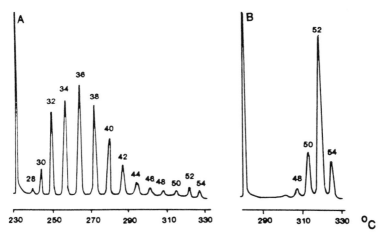

Figure 5.7 The separation of intact triacylglycerols of (A) coconut oil and (B) pig adipose tissue. Separation on a glass column (50 cm × 4 mm i.d.) packed with 1% SE-30™ on Chromosorb W™ (acid-washed and silanized; 100–120 mesh). Nitrogen at 50 ml/min was the carrier gas and for separation (A), the oven was temperature-programmed from 230 to 330°C at 2°C/min, while for separation (B) it was programmed from 280 to 330°C at 2°C/min. Reproduced with permission (Christie, 1989).

were non-polar. The columns used for this purpose are shorter than usual (50–100 cm, 2 mm id) and are packed with non-polar stationary phases such as SE30, OVI, Dexsil 30 which are stable up to 350°C. Elution is effected with a temperature programme from about 180 to 350°C. Under these conditions separation is according to the total number of carbon atoms in the three acyl chains (carbon number). It is interesting to consider what results are obtained in this way with palm kernel oil (a lauric oil rich in C_8–C_{18} acids) and with pig adipose tissue (almost entirely C_{16} and C_{18} acids (Figure 5.7).

The pig fat shows only four significant peaks of carbon number 48, 50, 52, and 54. The 48 peak will contain mainly tripalmitin while the remaining peaks will have one, two, or three C_{18} chains which will be stearic, oleic, or linoleic acid. Despite the limitations of these results the proportion of these four categories of triacylglycerols has provided valuable insight into the nature of oils rich in C_{16} and C_{18} acid including palm oil, cocoa butter, and many of its substitutes and extenders. The proportion of C_{50} and C_{52} glycerides (i.e. those having one or two C_{16} chains) is particularly important.

Because of the wider range of chain length (C_6–C_{18}, see sections 3.2.4 and 3.2.11) the gas chromatographic results for coconut oil and palm kernel oil seem more interesting with glycerides between C_{28} and C_{54}. Some typical results for these are listed in Table 5.4. On the basis of this method alone the content of individual glycerides cannot be determined. The number of individual glycerides with the same carbon number may be quite large. For example, 8,10,12 represents three different glycerides (ignoring stereoisomers) while 12,14,18 represents nine glycerides since 18 may be stearate, oleate, or linoleate.

With improved stationary phases it is now possible to separate triacylglycerols on more polar columns and also on capillary systems. Major separation by carbon number is now accompanied by a smaller separation depending on the number of double bonds. Symbols such as 54:1, 54:2, 54:3, etc. are used to distinguish C_{54} glycerides with a total of 1, 2, or 3 double bonds. A procedure with capillary (0.25 mm id) or wide-bore columns (0.53 mm id) has recently been described (Sassano and Jeffrey, 1993) (Figure 5.8).

This method can also be applied to diacylglycerols, phospholipids, and waxes. The latter contain only two long-chains and present no great difficulty. Phospholipids are converted to diacylglycerols by lipolysis with phospholipase C and these are examined as their trimethylsilyl derivatives (Figures 5.9 and 5.10).

(b) *High performance liquid chromatography.* Better results are usually obtained with HPLC using a reverse-phase system and this is the method of

Table 5.4 Glycerol esters (% wt) present in coconut oil and palm kernel oil as measured by gas chromatography

Carbon number	Coconut	Palm kernel	Major glycerides		
30	4.2	1.3	6,12,12	8,10,12	
32	15.8	8.6	6,12,14	8,12,12	
34	19.0	9.8	8,12,14	10,12,12	
36	20.3	25.7	12,12,12		
38	17.2	16.8	12,12,14		
40	9.6	9.5	12,12,16	12,14,14	
42	6.4	7.8	8,16,18	12,12,18	12,14,16
44	3.2	5.8	8,18,18	12,14,18	12,16,16
46	1.5	4.0	12,16,18	14,14,18	
other	2.8	10.7			

Adapted from *The Lipid Handbook*, 2nd edn (1994) p. 87.

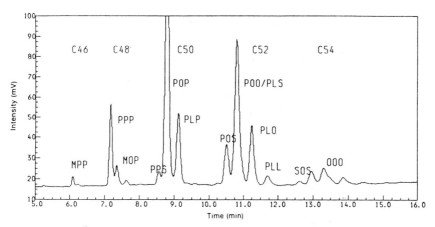

Figure 5.8 Separation of palm oil triacylglycerols by capillary gas chromatography. The column is a 15 m × 0.53 mm id methyl 65%-phenyl wide-bore capillary column; 345–365°C at 1°C/min. Abbreviations: M, P, O, S, L = myristic, palmitic, oleic, stearic and linoleic acids, respectively. (Reproduced with permission from G.J. Sassano and B.S.J. Jeffrey (1993).

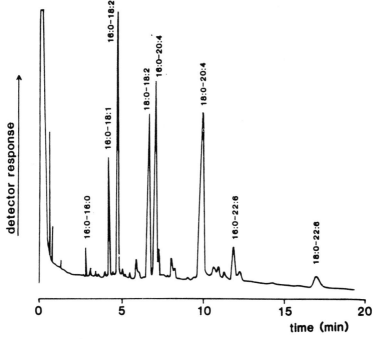

Figure 5.9 Separation by high temperature GLC of the trimethylsilyl ether derivatives of diacylglycerols, prepared by phospholipase C hydrolysis form the phosphatidylcholines of rat liver. The column was a 10 m × 0.25 mm glass capillary coated with SP-2330™, and was temperature-programmed from 190°C to 250°C at 20°C/min, then was held isothermally at 250°C. Splitless injection was used with hydrogen as the carrier gas. Only a few of the major peaks are identified here for illustrative purposes. Reproduced with permission (Christie, 1989).

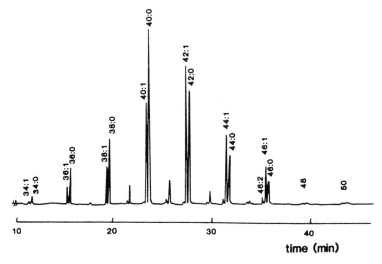

Figure 5.10 GC separation of the wax esters from the alga, *Chlorella kessleri*. A fused silica WCOT column (25 mm × 0.2 mm i.d.), coated with a methylsilicone phase, was temperature-programmed from 250 to 350°C at 2°C/min, with helium as the carrier gas. Reproduced with permission (Christie, 1989).

general choice. In reverse-phase HPLC the silica is coated with a non-polar phase (e.g. ODS octadecylsilica) and elution is effected with a non-polar solvent. Results obtained by a gradient elution system are given in Table 5.5.

Each glyceride category is associated with a carbon number corrected by the number of double bonds (db) and called a partition number. The partition number is related to the carbon number by the relationship:

$$PN = CN - 2db$$

The reversed phase HPLC system effects separation by partition number but the factor 2 in the above equation is only approximate and not always the same. For example, glycerol esters with one linoleate chain or two oleate chains (each with two double bonds), have the same partition number but may yet be separated. Thus the data in Table 5.5 shows six glyceride categories with a partition number of 48, namely OOO (54:3), LOS (54:3), OOP (52:2), LPS (52:2), OPP (50:1), and PPP (48:0). Only OOO and PPP are single components: OOP and OPP represent mixtures of two isomers and LOS and LPS represent mixtures of three isomers.

(c) *Silver ion systems.* Silver ions interact with double bonds and can be incorporated into chromatographic systems to produce another basis of separation. In the simplest case silver nitrate is incorporated onto a silica plate and under appropriate elution conditions it is possible to separate

Table 5.5 Triacylglycerol composition (% wt) by HPLC

Glyceride[a]	Carbon number	Partition number	Soybean	Olive	Safflower	Tallow
LLnLn	54:8	38	0.6	–	0.3	–
LLLn	54:7	40	7.8	–	–	–
LLL	54:6	42	22.5	–	36.1	–
LLnO	54:6	42	–	–	3.5	–
PLLn	52:5	42	2.9	–	0.4	–
LLO	54:5	44	18.2	1.1	21.5	–
LLP	52:4	44	14.6	0.7	17.7	–
LOO	54:4	46	9.5	14.8	5.2	–
LLS	54:4	46	3.5	–	5.4	–
LOP	52:3	46	9.4	3.7	5.3	0.3
LPP	50:2	46	1.5	–	0.4	1.6
OOO	54:3	48	3.0	44.6	1.1	2.2
LOS	54:3	48	2.4	–	1.2	–
OOP	52:2	48	2.0	23.3	0.4	16.0
LPS	52:2	48	0.7	–	0.2	3.1
OPP	50:1	48	0.2	2.3	0.2	12.7
PPP	48:0	48	–	0.6	–	5.9
OOS	54:2	50	0.4	4.5	0.1	7.6
OPS	52:1	50	–	0.5	0.1	18.9
PPS	50:0	50	–	0.1	–	6.1
OSS	54:1	52	–	–	–	6.7
PSS	52:0	52	–	–	–	4.4
SSS	54:0	54	–	–	–	0.9
other			0.8	3.8	0.9	13.6

Ln = linolenic, L = linoleic, O = oleic, S = stearic, P = palmitic.
[a]Each entry refers to all triacylglycerols with the three acids indicated.

triacylglycerols with 0–9 double bonds and sometimes even glycerides with the same total number of double bonds but made up in different ways, e.g. SMM from SSD (S = saturated, M = monoene, D = diene). Each separated fraction can be recovered from the plate, mixed with an internal standard (for purposes of quantitation), and converted to methyl esters for gas chromatography. An example is shown in Figure 5.11.

Alternatively a chromatographic column loaded with silver ions can be used in an HPLC system to separate glycerides mainly by double bond number. This is illustrated in an examination of palm oil (Figure 5.12). The peaks are well resolved but the saturated acids palmitic and stearic are not distinguished from each other.

Triacylglycerol mixtures can also be analysed by silver ion supercritical fluid chromatography.

5.2.5 Phospholipids

Phospholipids can be examined by any of the methods described above for triacylglycerols after suitable modification:

The phospholipid is submitted to lipolysis with phospholipase C and the resulting diacylglycerol is converted to an acetate, trifluoroacetate, or trimethylsilyl ether. In this form the sample can be analysed by gas chromatography or by HPLC.

Molecular species of individual phospholipids can also be separated by HPLC in reverse-phase mode.

5.2.6 Stereospecific processes

The processes described in the preceding sections give useful information about lipid composition but do little to associate acyl chains with the different glycerol carbon atoms. This problem has been tackled mainly, but not entirely, by enzymic procedures. These methods can be applied to individual lipid classes or to any subfraction of these.

(a) The first approach to this problem involved lipolysis with pancreatic lipase. This enzyme is 1,3-specific; it reacts only with acyl chains linked to

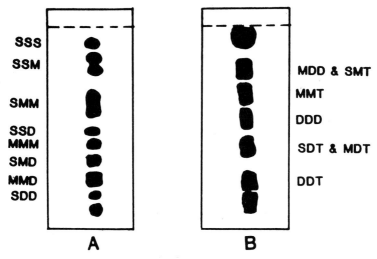

Figure 5.11 Schematic TLC separation of maize oil triacylglycerols on layers of silica gel G, impregnated with 10% silver nitrate. Mobile phases: Plate A, chloroform–methanol (99:1, v/v); Plate B, chloroform–methanol (96:4, v/v). Abbreviations: S, M, D and T denote saturated, mono-, di- and trienoic fatty acyl residues, respectively, esterified to glycerol. Reproduced with permission (Christie, 1989).

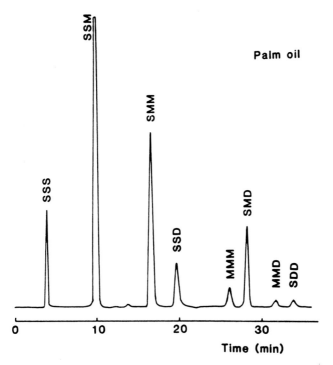

Figure 5.12 Silver ion HPLC separation of molecular species of triacylglycerols from palm oil on a column of Nucleosil 5SA^{TM} (250 × 4.6 mm) loaded with silver ions. The mobile phase was a linear gradient of 1,2-dichlorethane–dichoromethane (1:1, v/v) to this solvent with 50% acetone over 15 min, when acetonitrile was introduced to give a final mixture of acetone–acetonitrile (9:1, v/v) after a further 30 min, at a flow-rate of 0.75 ml/min and with mass detection. S = saturated, M = monoene, D = diene. Reproduced with permission (Christie, 1989).

the primary hydroxyl groups in glycerol, and under appropriate experimental conditions the following equilibrium is established:

There may be some further hydrolysis after isomerization of glycerol mono or diesters. After a reaction time of only a few minutes the total organic product is recovered (tri-, di-, and monoacylglycerols and free acids) and the 2-monoacylglycerols are isolated by thin-layer chromatography. These are converted to methyl esters and examined by gas chromatography. It is then possible to know the composition of the fatty acids at position *sn*-2 and, by difference, these in positions *sn*-1 and *sn*-3. This method does not allow a distinction to be made between the acids in these two positions.

| methyl esters from total oil/fat (% mol × 3) | 1, 2, 3 |
methyl esters from 2-monoacylglycerols (% mol)	2
difference (% mol × 2)	1, 3

The lipolysis equilibrium holds for a wide range of oils and fats and lipolytic study has been applied to many materials. The correctness of the results rely on the assumption that the lipase is 1,3-specific and also that it is not specific for fatty acids, i.e. it does not distinguish between fatty acids of differing structure at the α-positions. This is not wholly true. Short-chain acids react more quickly than the usual C_{16} and C_{18} acids, and acids with branched-chains or double bonds close to the ester function react more slowly. The method is therefore not entirely satisfactory with milk fats which contain short-chain acids, with fish oils containing acids with Δ4 unsaturation (DHA) and Δ5 unsaturation (EPA), and with seed oils containing petroselinic acid or γ-linolenic acid both of which are Δ6 acids. This, however, has not stopped oils of this kind being examined in this way.

Table 5.6 Component acids of some vegetable and animal fats and the 2-monoacylglycerols derived from them by lipolysis

Source	16:0[a]	18:0	18:1	18:2	Other[b]
rape (high-erucic)	3	–	16	10	18:3 (14, 31) 20:1 (9, 0)
	1	–	41	26	22:1 (47, 1)
cocoa butter	27	33	35	3	
	2	2	85	11	
palm oil	44	6	39	9	
	11	2	65	22	
poppy	10	2	11	76	
	1	–	9	89	
olive	12	2	76	8	
	1	–	88	10	
pig	28	15	42	9	16:1 (3, 4)
	72	4	12	3	
sheep	27	27	35	2	16:1 (3, 5)
	14	9	58	6	
ox	30	25	36	1	16:1 (5, 6)
	14	8	61	3	

[a]The two rows of figures refer to triacylglycerols (*sn*-1, 2, and 3) and 2-monoacylglycerols (*sn*-2), respectively.
[b]The two figures given in parenthesis refer to triacylglycerols and 2-monoacylglycerols, respectively.

This method quickly and clearly demonstrated that fatty acids are not generally distributed at random. The fatty acid composition of the 2-monoacylglycerols is different from that of the triacylglycerols from which they are derived. In most vegetable oils the 2-position is acylated almost entirely by unsaturated C_{18} acids, while saturated acids and longer-chain unsaturated acids such as erucic (when present) are found in the α-positions along with unsaturated C_{18} acid not accommodated in the 2-position. Some typical results are collected together in Table 5.6. Results with animal fats are less uniform and pig fat is unusual in the high level of palmitic acid in the 2-positions. A similar result holds for human milk fat.

Results obtained in this way can be converted into glyceride composition of the traditional kind only if two assumptions are made: (i) that the acids at C1 and C3 are the same; and (ii) that the acids at C1, C2, and C3 are randomly associated with each other. Even if this is true for oils and fats (see later) it will not be true for subfractions of the whole sample.

(b) Phospholipids can be examined by a slight modification of this procedure. Hydrolysis with phospholipase A_2 (available from snake venom) liberates fatty acids from the 2-position and leaves a lysophospho-lipid:

phospholipid   lysophospholipid
(PZ = phosphate ester group)

It is then possible to find what fatty acids are present at each of positions 1 and 2. Examples of some results are given in Table 5.7. In general, saturated acids are found in position 1 and unsaturated acids in position 2 but despite this tendency phospholipids with two saturated acids or two unsaturated acids are also known.

(c) A procedure which distinguishes fatty acids at all three glycerol carbon atoms was developed by Brockerhoff (1971) and improved by Myher and Kuksis (1979). This depends on incomplete and random reaction of triacylglycerols with a Grignard reagent (EtMgBr), isolation of the two α,β-diacylglycerols and separation from the αα-compounds, and conversion of these to a phospholipid or some other phosphate ester. Finally, reaction of these products with phospholipase C will distinguish between the two αβ isomers since only one of the pair has the same stereochemistry as a natural phospholipid. These procedures are outlined in Figures 5.13 and 5.14 and some typical results are collected together in Table 5.8.

Table 5.7 The distribution of major component acids (% wt) in some phosphatidylcholines

Source	16:0[a]	18:0	18:1	18:2	20:4	20:5	22:6
salmon	37	3	8	–	–	14	33
	10	0	23	–	–	17	46
egg	61	25	10	2	0	–	–
	5	2	59	26	6	–	–
rat liver	32	50	4	5	5	4	0
	2	1	7	24	45	3	14
bovine milk[b]	47	20	23	3	–	–	–
	35	3	32	11	–	–	–

[a]The two rows of figures refer to fatty acids in the *sn*-1 and *sn*-2 positions, respectively.
[b]Also 14:0 (6, 13).

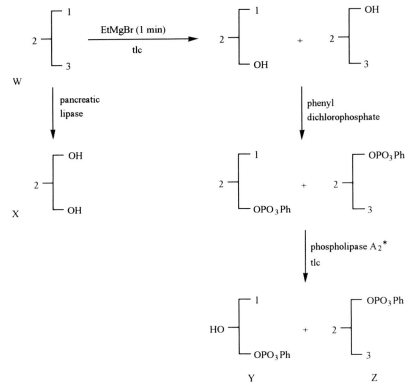

Figure 5.13 Stereospecific analysis of triacylglycerols. 1, 2, 3 represent fatty acids at the *sn*-1, *sn*-2, and *sn*-3 positions, respectively. The fatty acids present in fractions W to Z are determined by chromatography of the methyl esters: *sn*-1 = Y *sn*-2 = X *sn*-3 = 3W-X-Y or 2Z-X where X to Z are fatty acid compositions (% mol). If the experiment is satisfactory the C3 values calculated in two ways will show reasonable agreement. (Brockerhoff (1971). *Only the phenylphosphate with the correct stereochemistry reacts with phospholipase A$_2$.

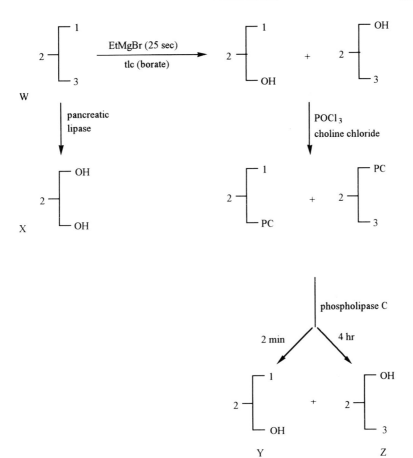

Figure 5.14 Stereospecific analysis of triacylglycerols. 1, 2, 3 represent fatty acids at the *sn*-1, *sn*-2, and *sn*-3 positions respectively. (i) Examine products W to Z as methyl esters for stereospecific analysis: *sn*-1 = 2Y-X or 3W-2Z, *sn*-3 = 3W-2Y or 2Z-X. (ii) Examine Y and Z as tertbutyldimethylsilyl ethers for information on molecular association. (Myher and Kuksis (1979)).

In the Brockerhoff procedure the acids at C3 were not measured directly but were derived by calculation in two ways. They should agree to within 3%. The Myher–Kuksis method which produces actual phospholipids rather than structures resembling phospholipids gives more accurate results. Also by chromatographic examination of derivatives of the diacylglycerols Y and Z it is possible to discover how the acids are paired in each of these compounds.

(d) Christie and Takagi have each developed processes of stereospecific analysis based on urethanes, which do not involve the use of enzymes.

Table 5.8 Stereospecific analyses of some natural fats

Source		16:0[a]	18:0	18:1	18:2	Other[b]
soybean	1	14	6	23	48	18:3 (9, 7, 8)
	2	1	tr	22	70	
	3	13	6	28	45	
cocoa butter	1	34	50	12	1	
	2	2	2	87	9	
	3	37	53	10	tr	
rape (high-erucic)	1	4	2	23	11	18:3 (6, 20, 3)
	2	1	0	37	36	20:1 (16, 2, 17)
	3	4	3	17	4	22:1 (35, 4, 51)
pig	1	16	21	44	12	
	2	59	3	17	8	
	3	2	10	64	24	
ox	1	41	17	20	4	14:0 (4, 9, 1)
	2	17	9	41	5	16:1 (6, 6, 6)
	3	22	24	37	5	
chicken	1	25	6	33	14	16:1 (12, 7, 12)
	2	15	4	43	23	
	3	24	6	35	14	

[a]The three rows of figures show fatty acid composition in positions sn-1, 2, and 3, respectively.
[b]The three figures in parenthesis indicate the percentage of acid at the sn-1, 2, and 3 positions, respectively.

In Christie's procedure triacylglycerols are reacted briefly with EtMgBr and the reaction product is converted by reaction of all free hydroxyl groups with S-(+)-1-(1-naphthyl)ethyl isocyanate (Figure 5.15). The three types of diacylglycerol urethanes are separated from other products by chromatography on solid phase extraction columns and then from one another by HPLC. The urethanes based on 1,3-diacylglycerols are separated first followed by those based on 1,2- and 2,3-diacylglycerols. These latter are separable because the two kinds of urethanes are diastereoisomeric. In the following equation A represents the enantiomeric isocyanate and B the racemix mixture of diacylglycerols. The two urethanes are diastereoisomeric and will thus have different physical properties.

$$(S)\text{-}A + (R,S)\text{-}B \rightarrow (S)\text{-}A\text{-}(R)\text{-}B + (S)\text{-}A\text{-}(S)\text{-}B$$
$$\text{urethanes}$$

The separated urethanes are converted to methyl esters for gas chromatographic examination and the fatty acid composition at each position is determined either from the above data or by measuring fatty acids at sn-2 by pancreatic lipolysis:

sn-1 from TAG and 2,3-DAG or from 1,2-DAG and 2-MAG
sn-2 from TAG, from *sn*-1 and *sn*-3, or from 2-MAG
sn-3 from TAG and 1,2-DAG or from 2,-3-DAG and 2-MAG

Typical results for palm oil and for tallow are given in Table 5.9.

Takagi studies the monoacylglycerols produced from triacylglycerols by reaction with EtMgBr and separates the 2-monoacyl esters from the enantiomeric 1- and 3-monoacyl esters. The latter are converted to urethanes by reaction with 3,5-dinitrophenylisocyanate and the enantiomers are separated on a chiral column. Each monoacylglycerol group is then converted to methyl esters for gas chromatography (Figure 5.16). Some typical results are given in Table 5.10.

5.2.7 Combined processes

The various methods which have now been described provide useful insight into lipid composition. It is, however, difficult to combine some of these

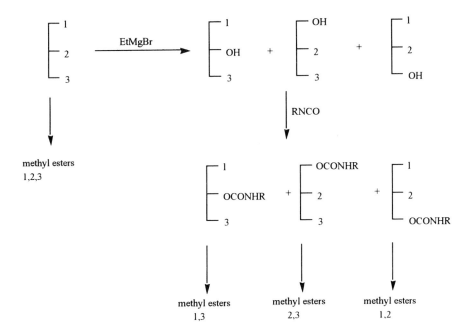

RNCO = NpCH(CH$_3$)NCO (Np= 1-naphthyl)

Figure 5.15 Stereospecific analysis of triacylglycerols using urethanes of diacylglycerols. 1, 2, 3 represent fatty acids at the *sn*-1, *sn*-2, and *sn*-3 positions, respectively. (Laakso and Christie (1991)).

Table 5.9 Fatty acid composition (% mol) of triacylglycerols and of positions *sn*-1, *sn*-2, and *sn*-3

Source	TAG	*sn*-1	*sn*-2	*sn*-3
Palm oil[a]				
16:0	48.4	60.1	13.3	71.9
18:0	3.7	3.4	0.2	7.6
18:1	36.3	26.8	67.9	14.4
18:2	10.0	9.3	17.5	3.2
Tallow[a]				
14:0	3.4	2.9	1.5	5.7
16:0	29.5	42.0	24.6	21.9
18:0	26.0	34.4	11.3	32.4
18:1	33.1	13.6	55.0	30.8

[a]The original paper includes minor component acids also.
(W.W. Christie, B. Nikolova-Damyanova, P. Laakso, and B. Herslof (1991), *J. Amer. Oil Chem. Soc.*, **68**, 695).

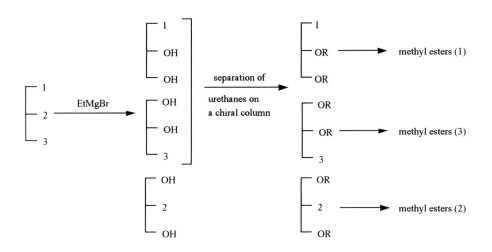

R= CONHR'   (R' = 3,5-dinitrophenyl)

Figure 5.16 Stereospecific analysis of triacylglycerols using urethanes of monoacylglycerols. 1, 2, 3 represent fatty acids at the *sn*-1, *sn*-2, and *sn*-3 positions, respectively. (Itabashi and Takagi (1986)).

results and this can only be achieved when they are carried out in a sequential manner. Quantity of sample then becomes important and the following should be noted:

1. Separation by silver ion chromatography on plates or columns can usually be carried out on a scale which will allow further examination by

Table 5.10 Fatty acid composition (% mol) of triacylglycerols and of positions *sn*-1, *sn*-2, and *sn*-3

Source	TAG	*sn*-1	*sn*-2	*sn*-3
Rapeseed oil[a]				
16:0	3.9	5.6	1.6	4.5
18:1 (*n*-9)	17.2	13.7	32.4	5.1
18:2	16.1	1.2	39.3	7.4
18:3	8.7	3.3	21.3	1.3
20:1 (*n*-9)	8.9	15.5	1.5	10.0
22:1 (*n*-9)	37.4	50.2	0.8	61.3
Herring oil[a]				
14:0	9.7	11.9	6.7	9.7
16:0	15.7	15.2	17.9	12.5
16:1 (*n*-7)	5.0	6.4	3.8	4.3
18:1 (*n*-9)	9.9	11.3	8.7	9.0
20:1 (*n*-9)	11.8	16.3	4.2	14.4
20:5 (*n*-3)	5.9	3.0	10.5	3.3
22:1 (*n*-11)	16.0	12.3	5.2	31.6
22:6 (*n*-3)	6.1	2.2	14.9	–

[a]The original paper included minor components also.
(T. Takagi and Y. Ando (1991), *Lipids*, **27**, 542).

a second procedure (e.g. gas chromatography of methyl esters or stereospecific analyses).

2. HPLC can also give fractions for further examinations by chromatographic or spectroscopic procedures.
3. Since the stereospecific procedure breaks down the original molecules this will limit further investigation.
4. Gas chromatography requires only small amounts of material and when carried out on methyl esters must come at the end of a reaction sequence. Sometimes gas chromatography is carried out in a preparative mode. It can also be coupled with mass spectrometry.

The most common combined procedures involve the use of silver ion chromatography (plate or column) followed by further examination of each fraction by gas chromatography or HPLC. In the latter case fractions can be collected and further converted to methyl esters for gas chromatographic investigation. The silver ion system effects separation by double bond index (unsaturation) and the subsequent GLC or HPLC system separates each fraction by carbon number (total carbon atoms in the three acyl chains).

An example of this is a study of herring oil (Laakso and Christie, 1991). The oil, with a double bond index of 4.9, was separated by silver ion HPLC into 11 fractions of double bond index 0–9.6. Each fraction was then separated further by reverse-phase HPLC and a total of 130 fractions were obtained. Each of these was converted to methyl esters and examined by gas chromatography. Thus for each fraction there is a double bond index, a carbon number, and a methyl ester composition. The results are too extensive to reproduce but a comment is made about the largest 'single component'. This comes from Ag^+ fraction ten with double bond index of 6.9 and is a single HPLC fraction consisting mainly of 16:0 30.3% mol (and 2.3% of other saturated acids), 22:1 29.1% (and 3.6% of other monoene acids), and 22:6 32.8% (and 0.1% of other polyene acids). Most of this fraction must therefore consist of triacylglycerols containing one mole of 16:0, 22:1, and 22:6. There will be six such triacylglycerols without allowance for the fact that 22:1 may represent more than one isomer.

Bibliography

Bligh, E.G. and W.J. Dyer (1959) A rapid method of total lipide extraction and purification, *Can. J. Biochem. Physiol.*, **37**, 911–917.

Blomberg, L.G. and M. Demibuker (1994) Analysis of triacylglycerols by argentation supercritical fluid chromatography, in *Developments in the Analysis of Lipids* (eds J.H.P. Tyman and M.H. Gordon), Royal Society of Chemistry, Cambridge.

Brockerhoff, H. (1971) Stereospecific analysis of triglycerides, *Lipids*, **6**, 942–956.

Christie, W.W. (1987) *High-performance Liquid Chromatography and Lipids*, Pergamon, Oxford.

Christie, W.W. (1987) A stable silver-loaded column for the separation of lipids by high-performance liquid chromatography, *J. High Res. Chromatogr. Commun.*, **10**, 148–150.

Christie, W.W. (1989) *Gas Chromatography and Lipids*, The Oily Press, Ayr.

Christie, W.W. (1989) Silver ion chromatography using solid-phase extraction columns packed with a bonded-sulphonic acid phase, *J. Lipid Res.*, **30**, 1471–1473.

Christie, W.W. (ed.) (1992) *Advances in Lipid Methodology – One*, The Oily Press, Ayr.

Christie, W.W. (ed.) (1993) *Advances in Lipid Methodology – Two*, The Oily Press, Dundee.

Christie, W.W. (1994) High performance liquid chromatography of chiral lipids, in *Developments in the Analysis of Lipids* (eds J.H.P. Tyman and M.H. Gordon) Royal Society of Chemistry, Cambridge.

Folch, J., M. Lees and G.H.S. Stanley (1975) A simple method for the isolation and purification of total lipides from animal tissue, *J. Biol. Chem.*, **226**, 497–509.

Gunstone, F.D. *et al.* (eds) (1994) *The Lipid Handbook*, 2nd edn, Chapman and Hall, London.

Hamilton, R.J. and J.B. Rossell (eds) (1986) *Analysis of Oils and Fats*, Elsevier Applied Science, London.

Hammond, E. (1993) *Chromatography for the Analysis of Lipids*, CRC, Boca Raton.

Itabashi, Y. and T. Takagi (1986) High performance liquid chromatographic separation of monoacylglycerol enantiomers of a chiral stationary phase, *Lipids*, **21**, 413–416.

Laakso, P. and W.W. Christie (1990) Chromatographic resolution of chiral diacylglycerol derivatives: potential in the stereospecific analysis of triacyl-*sn*-glycerols, *Lipids*, **25**, 349–353.

Laakso, P. and W.W. Christie (1991) Combination of silver ion and reversed-phase high-performance liquid chromatography in the fractionation of herring oil triacylglycerols, *J. Amer. Oil Chem. Soc.*, **68**, 213–223.

Myher, J.J. and A. Kuksis (1979) Stereospecific analysis of triacylglycerols via racemic phosphatidylcholines and phospholipase C, *Can. J. Biochem.*, **57**, 117–124.

Robbelen, G. *et al.* (eds) (1989) *Oil Crops of the World – Their Breeding and Utilisation*, McGraw-Hill, New York.

Rossell, J.B. and J.L.R. Pritchard (eds) (1991) *Analysis of Oilseeds, Fats, and Fatty Foods*, Elsevier Applied Science, London.

Sassano, G.J. and Jeffrey, B.S.J. (1993) *J. Amer. Oil Chem Soc.*, **70**, 1111–1114.

6 Physical properties

6.1 Polymorphism, crystal structure, and melting point

6.1.1 Introduction

In the solid state long-chain compounds frequently exist in more than one crystalline form and may consequently have more than one melting point. This property of polymorphism is of both scientific and technical interest. The understanding of this phenomenon is essential for the satisfactory blending and tempering of fat-containing materials such as cooking fats and confectionery which must attain a certain physical appearance during preparation and maintain this during storage. Problems of graininess in margarine and bloom in chocolate for example, are both related to polymorphic changes.

The experimental methods used most extensively to examine melting and crystallization behaviour involve dilatometry and low-resolution ^{1}H NMR spectroscopy, differential scanning calorimetry (DSC), infrared spectroscopy, and X-ray diffraction (XRD). Polymorphism is apparent through the melting behaviour of individual compounds. It has sometimes been difficult to correlate the observations made in these different ways and the problem has been compounded by the difficulty – at least in the early studies – of obtaining pure compounds (especially mixed glycerol esters) for examination and by the fact that natural fats are complex mixtures.

X-ray investigations indicate that the unit cell for long-chain compounds is a prism with two short spacings and one long spacing as indicated in Figure 6.1. When the long-spacing is less than the molecular dimension calculated from known bond lengths and bond angles, it is assumed that the molecule is tilted with respect to its end planes. On the other hand, the length may be such as to indicate a dimeric or a trimeric unit. The molecules tend to assume the angle of tilt at which they are most closely packed, when they will have the greatest stability and the highest melting point. In the following discussion acids (and the glycerol esters) with an even and an odd number of carbon atoms in the acyl chain(s) are described as even and odd, respectively.

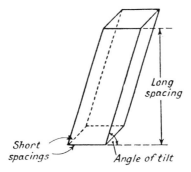

Figure 6.1 The unit cell for long-chain compounds.

6.1.2 Alkanoic acids

The melting points of long-chain acids and their methyl esters are listed in Table 1.1. These values show 'alternation' with increasing chain length, a phenomenon commonly displayed by the physical properties of long-chain compounds in the solid state and related to the arrangements of molecules within the crystals. The melting points of the even acids and their methyl esters plotted against chain length fall on smooth curves lying above similar curves for the odd acids and their methyl esters. Odd acids melt lower than even acids with one less carbon atom. The two curves for saturated acids converge at 120–125°C.

The melting points of unsaturated acids depend not only on chain length but on the nature of the unsaturation (*cis*- or *trans*-olefinic or acetylenic) and on the number and relative position of unsaturated centres. For example stearic (70°C), oleic (Δ9c, 11°C), elaidic (Δ9t, 45°C) and stearolic acids (Δ9a, 46°C) have the melting points shown. Among 31 isomeric 18:1 acids both *cis*- and *trans*-isomers show alternation (with respect to double bond position) with the *trans*-isomers being higher melting (Table 6.1). Similar results have been reported with the glycerol esters of these octadecenoic acids (Figure 6.2).

Among polyunsaturated acids those with conjugated unsaturation are higher melting than their methylene-interrupted isomers. Some examples are given in Table 6.2.

The alkanoic acids exist in three crystalline forms designated A, B, and C for even acids and A′, B′, and C′ for odd acids. These sets are in order of decreasing angle of tilt (Table 6.3).

The relationship between these three crystalline forms and the liquid melt is shown below. In the even acids only form C is obtained from the melt. This is the most stable (highest melting) form and has the melting point which has been given in Table 1.1. Crystallization from polar solvents usually gives C but from non-polar solvents either form A or forms B and C

Table 6.1 Melting points (°C) of octadecenoic acids and their glycerol esters and of octadecynoic acids

Position of unsaturation	Acids			Glycerol esters	
	acetylenic	*cis*	*trans*	*cis*	*trans*
2	57	50	58	–	–
3	74	50	65	–	–
4	75	46	59	34	53
5	52	13	47	14	41
6	51	29	54	28	52
7	49	13	45	7	39
8	47	24	52	24	49
9	46	11	45	5	41
10	46	23	53	27	49
11	47	13	44	10	43
12	47	28	53	32	51
13	49	27	44	26	44
14	64	42	53	44	58
15	65	41	59	43	56
16	72	54	66	–	–
17	67		56		55

Melting points of stearic acid and glycerol tristearate are 70°C and 73°C, respectively.
J.A. Barve and F.D. Gunstone (1971), *Chem. Phys. Lipids.*, **7**, 311.
J.W. Hagemann, W.H. Tallent, J.A. Barve, I.A. Ismail, and F.D. Gunstone (1975), *J. Amer. Oil Chem. Soc.*, **52**, 204.

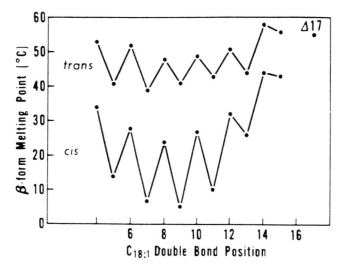

Figure 6.2 Melting points of 18:1 glycerol triesters (β form) versus double bond position. Reproduced with permission (Hagemann *et al.* (1975)).

Table 6.2 Melting points (°C) of some monounsaturated and polyunsaturated acids

monoene	
16:1 (9c)	0.5
18:1 (9c)	16.3 and 13.4
20:1 (9c)	25
22:1 (13c)	33.4
polyenes with methylene-interrupted unsaturation	
18:2 (9c12c)	−5
18:2 (9c12t)	−3
18:2 (9t12t)	29
18:3 (9c12c15c)	−11
18:3 (9t12t15t)	30
20:4 (5c8c11c14c)	−49.5
polyenes with conjugated unsaturation	
18:2 (9c11t)	22
18:2 (9t11t)	54
18:3 (9c11t13c)	44
18:3 (9c11t13t)	49
18:3 (9t11t13c)	32
18:3 (9t11t13t)	71

Table 6.3 Crystallographic data for straight-chain alkanoic acids

Acid type	Crystal form	Symmetry of the unit cell	Moles per unit cell	Angle of tilt
even acids				
	A	triclinic	2	67
	B	monoclinic	4	66
	C[a]	monoclinic	4	56
odd acids				
	A′	triclinic	2	66
	B′[a]	triclinic	4	61
	C′	monoclinic	4	59

[a]The most stable (highest melting) crystalline form.

are obtained. Among odd acids B′ is the most stable form and crystallizes from the melt.

even acids   odd acids

The molecules crystallize in dimeric layers as indicated in the simplified representation in Figure 6.3. Alternation results from the fact that the methyl groups in the end group plane interact differently in the odd and even series.

6.1.3   Glycerol esters

The important triacylglycerols are discussed first, with some brief reference to monoacylglycerols and diacylglycerols at the end.

It has been known, since the classical work of Chevreul in the 1820s, that fats, unlike most other organic compounds, show multiple melting point. As far back as 1853 glycerol tristearate was reported to have three melting points (52°C, 64°C, 70°C). Understanding this phenomenon proved difficult because of inadequate experimental techniques, a number of false leads, and the difficulty of obtaining pure compounds for study. Successful investigations started with simple saturated triacylglycerols of the type GA_3 and only later were mixed glycerides and unsaturated compounds examined.

X-ray powder diffraction studies confirmed the existence of three crystalline forms with different short and long spacings but, at first, these were incorrectly correlated with the melting points. This difficulty was overcome by the study of their infrared spectra and important data are summarized in Table 6.4.

When the melt of a simple triacylglycerol is cooled quickly it solidifies in the lowest melting form (α) which has perpendicular alkyl chains (i.e. the angle of tilt is 90°). When heated slowly this melts and, held just above the α melting point, it will resolidify in the β' crystalline form. Similarly a more stable β form can be obtained from the β' form. The β form has the highest melting form and is produced also by crystallization from solvent.

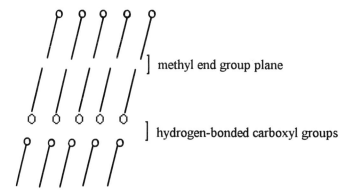

Figure 6.3 Schematic arrangement of alkanoic acid molecules in the crystalline form. o represents the polar head group (COOH), and the line represents the alkyl chain which will assume a zig-zag arrangement of successive carbon atoms.

Table 6.4 Characterization of α, β', and β forms of crystalline triacylglycerols

mp	Short spacings (nm)	Infrared absorption (cm^{-1})	Hydrocarbon chain	Subcell
α lowest	0.4	720	perpendicular	orthorhombic
β' intermediate	$\begin{cases} 0.42\text{--}0.43 \\ \text{and} \\ 0.37\text{--}0.40 \end{cases}$	$\left.\begin{array}{l} 726 \\ \\ 719 \end{array}\right\}$	tilted	orthorhombic
β highest	$\begin{cases} 0.46 \text{ and} \\ 0.36\text{--}0.39 \end{cases}$	717	tilted	triclinic

These transformations are not reversible. This series of changes is shown by the sequence:

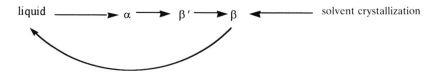

liquid α β' β solvent crystallization

Data on the GA_3 glycerol esters (C_8–C_{22}) are summarized in Table 6.5a and b. These show the marked differences in melting point and long-spacing for the three crystalline forms. For example, for trilaurin these are $15.0°$ (35.6×10^{-10} m), $34.5°$ (32.9×10^{-10} m), and $46.5°$ (31.2×10^{-10} m) for the α, β' and β form.

With mixed saturated triacylglycerols such as PStP (P = palmitic, St = stearic) the β form is only obtained with difficulty and such compounds have a highest-melting β' form. Among unsaturated glycerol esters symmetrical compounds (SUS, USU, UUU) usually have highest-melting β forms while unsymmetrical compounds (USS, USS) have a stable β' form. (In these formulations S = saturated and U = unsaturated.)

The stable β form generally crystallizes in a double chain length arrangement (DCL, or $\beta2$) but if one acyl group is very different from the other two in chain length or in degree of unsaturation the crystals assume a triple chain length arrangement (TCL, $\beta3$) since this permits more efficient packing of alkyl chains and head groups. These crystals have the short spacing expected of a β crystalline form but the long spacing is about 50% longer than usual.

In the DCL arrangement the molecules align themselves (like tuning forks) with the two chains in extended line (to give the double chain length) and the third chain parallel to these (Figure 6.4). Some mixed glycerides (GA_2B) which have a TCL form when crystallized on their own, give high-melting (well-packed) mixed crystals with a second appropriate glyceride (e.g. CPC and PCP or OPO and POP, C = capric, P = palmitic, O = oleic) (Figure 6.4). This has been described as 'compound formation'.

Table 6.5a Melting points and long spacings of mono-acid triacylglycerols

Acid chain length	Melting point (°C)			Long spacing ($\times 10^{-10}$ m)		
	α	β'	β	α	β'	β
8	−51	−18.0	10.0	–	–	22.7
9	−26	4.0	10.5	–	25.3	24.9
10	−10.5	17.0	32.0	30.2	27.7	26.5
11	2.5	27.0	31.0	32.7	29.8	29.6
12	15.0	34.5	46.5	35.6	32.9	31.2
13	24.5	41.5	44.5	37.8	34.2	34.0
14	33.0	46.0	58.0	41.0	37.3	35.7
15	39.0	51.5	55.0	42.9	39.2	39.2
16	45.0	56.5	66.0	45.8	42.5	40.8
17	50.0	60.5	64.0	48.5	43.8	43.5
18	54.7	64.0	73.3	50.6	47.0	45.1
19	59.0	65.5	71.0	53.1	48.1	48.2
20	62.0	69.0	78.0	55.8	50.7	49.5
21	65.0	71.0	76.0	58.5	53.2	52.7
22	68.0	74.0	82.5	61.5	56.0	54.0

Selected from: E.S. Lutton and A.J. Fehl (1970), *Lipids*, **5**, 90.

Table 6.5b Melting points (°C) of some mono-acid triacylglycerols

	α	β'	β
10:0	−10	13,18,26	33
12:0	14	30,34,40	46
14:0	31	41,45,51	56
16:0	46	53,57	66
18:0	55	61,64	73
20:0	64	69,71	78
22:0	69,70	74,77	83

J.W. Hagemann and J.A. Rothfus (1983), *J. Amer. Oil Chem. Soc.*, **60**, 1123.

The methyl groups at the top and bottom of each glyceride layer do not usually lie on a straight line, but form a boundary of a particular structure depending on the lengths of the several acyl groups. This is called the 'methyl terrace'. The glycerides tilt with respect to their methyl end planes to give the best fit of the upper methyl terrace of one row of glycerol esters with the lower methyl terrace of the next row of esters. There may therefore be several possible $\beta 2$ modifications differing in the slope of the methyl terrace and in the angle of tilt.

Among monoacylglycerols the 2-isomer shows only one crystalline form (β) but the 1-isomers are more complex as indicated by the scheme:

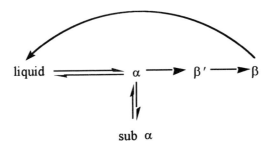

1,2-diacylglycerols exist in the forms shown below. The β' form is able to stabilize the β' form of triacylglycerols and so inhibits the conversion of the latter to the β form. 1,3-diacylglycerols form two β-type crystals.

1,2-diacylglycerols   liquid ⟶ α ⟶ β'

1,3-diacylglycerols   liquid ⟶ β_2 ⟶ β_1

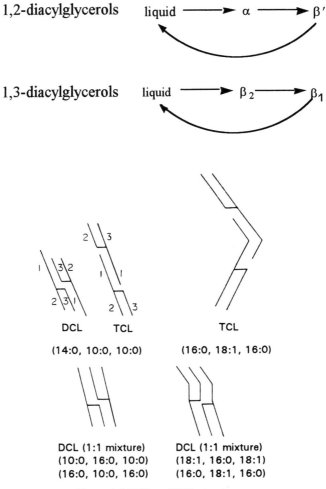

DCL   TCL   TCL
(14:0, 10:0, 10:0)   (16:0, 18:1, 16:0)

DCL (1:1 mixture)   DCL (1:1 mixture)
(10:0, 16:0, 10:0)   (18:1, 16:0, 18:1)
(16:0, 10:0, 16:0)   (16:0, 18:1, 16:0)

Figure 6.4 DCL and TCL structures.

6.1.4 Margarines and confectionery fats

In the production of margarines and shortenings the β' crystalline form is preferred to the β form. The β' crystals are relatively small and can incorporate a large amount of liquid. This gives the product a glossy surface and a smooth texture. The β crystals, on the other hand, though initially small, grow into needle-like agglomerates. These are less able to incorporate liquids and produce a grainy texture. Margarine and shortenings made from canola (rapeseed), sunflower, and soybean oil after partial hydrogenation tend to develop β crystals but this can be inhibited or prevented by the incorporation of some hydrogenated palm oil or palm olein which stabilize crystals in the β' crystal form.

Because of the importance of its melting behaviour the polymorphism displayed by cocoa butter has been thoroughly investigated. This material is particularly rich in three 2-oleo-1,3-disaturated glycerols namely POP, POSt, and StOSt. The solid fat has been identified in six crystalline forms designated I–VI with the melting points and DCL or TCL nature indicated in Table 6.6. Of these, form $V(\beta_2)$ is the one preferred for chocolate. This crystalline form gives good demoulding characteristics and has a stable gloss and a favourable snap at room temperature. Two procedures have been employed to promote the formation of this particular crystalline form. The most extensively used is tempering, i.e. putting molten chocolate through a series of cooling and heating processes which have been found to optimize the production of the appropriate polymorph. An alternative procedure requires seeding of the molten chocolate with cocoa butter already prepared in form $V(\beta_2)$ or form $VI(\beta_1)$. This latter method has been restricted by the difficulty of obtaining adequate supplies of these crystalline forms.

The synthetic glycerol ester 2-oleo 1,3-dibehenin (BOB) may be added to cocoa butter to prevent bloom formation by keeping it in its form V at temperatures above 30°C. The melting points of some 2-oleo disaturated glycerides and the chain length of their crystalline forms are summarized in Table 6.7.

6.2 Ultraviolet spectroscopy

The use of ultraviolet spectroscopy in the study of lipids is confined to systems containing or generating conjugated unsaturation. It is therefore of value in the study of natural acids with conjugated unsaturation such as the dienes, trienes, and acetylenic compounds described in sections 1.6, 1.8, and 1.9. Conjugated dienes have a maximum around 230–240 nm and trienes show a triple peak around 261, 271, and 281 nm. Methylene-interrupted polyenes undergo double bond migration to produce compounds with conjugated unsaturation in reactions such as oxidation or

Table 6.6 Polymorphism in cocoa butter

Parameter	I	II	III	IV	V	VI
mp (°C)	17.3	23.3	25.5	27.3	33.8	36.3
chain length[a]	D	D	D	D	T	T

[a]D = DCL; T = TCL

Table 6.7 Melting points of polymorphic forms of some 2-oleo 1,3-disaturated glycerol esters

Type	POP	StOSt	AOA	BOB
α	15.2[a]	23.5[a]	31.5[a]	41.5[a]
γ	27.0	35.4	45.5	49.5
pseudo β'	$\begin{Bmatrix}30.3\\33.6\end{Bmatrix}$[a]	36.5	46.5	50.5
β_2	35.1	41.0	46.6	53.0
β_1	36.7	43.0	48.3	53.3

[a]All polymorphs are TCL except for the DCL forms marked [a].
P = palmitic(16:0), O = oleic(18:1), S = stearic(18.0), A = arachidic(20:0). B = behenic (22.0).
I. Machiya, T. Koyeno, and K. Sato, (1990), *Lipid Technol.*, **2**, 34.

hydrogenation (sections 7.1 and 7.2) and the appearance of absorption at appropriate wavelength has been employed in the study of such processes.

6.3 Infrared spectroscopy

Infrared spectroscopy has been applied to solid lipids to provide information about polymorphism, crystal structure, conformation, and chain length but the commonest use of traditional IR spectroscopy has been the recognition and determination of *trans* unsaturation using liquids or solutions. One *trans* double bond gives characteristic absorption at 968 cm^{-1} and the frequency does not change for additional double bonds unless these are conjugated, when there are small changes from this value.

There is no similarly diagnostic infrared absorption bond for *cis* unsaturation but Raman spectra show strong absorption bonds at 1665 ± 1 cm^{-1} (*cis*-olefin), 1670 ± 1 cm^{-1} (*trans*-olefin), and 2230 ± 1 and 2291 ± 2 cm^{-1} (acetylenes) for the type of unsaturation shown.

Carbonyl compounds have a strong absorption band in the region 1650–1750 cm^{-1}. The wavelength varies slightly with the nature of the carbonyl compound (Table 6.8) and this may be of diagnostic value.

Most oils containing the usual mixture of saturated and unsaturated acids have similar infrared spectra. Superimposed on this there may be

Table 6.8 Infrared absorption bands (cm^{-1}) for saturated and $\alpha\beta$-unsaturated carbonyl compounds

Compound	Saturated	$\alpha\beta$-unsaturated
aldehydes	1740–1720	1705–1680
ketones	1725–1705	1685–1665
acids	1725–1700	1715–1690
esters	1750–1730	1730–1715
anhydrides	1850–1800 and 1790–1740	–
amides (1 ary)	1690–1650	–
amides (2 ary)	1680–1630	–
amides (3 ary)	1670–1630	–

additional absorption associated with less common functional groups such as hydroxyl (3448 cm^{-1}), keto (1724 cm^{-1}), cyclopropene (1852 and 1010 cm^{-1}), epoxide (848 and 826 cm^{-1}), allene (2222 and 1961 cm^{-1}), vinyl (990 and 909 cm^{-1}), and conjugated enyne (952 cm^{-1}).

The use of Fourier transform infrared spectrometers (FTIR) leads to improvements in wavelength control and absorption measurement. Small differences can be detected and exploited by techniques such as subtracting and ratioing. Based on these advantages methods have been developed for measuring iodine value, saponification value, free acid, and oxidative stability. These parameters can be measured by simple menu-driven procedures in a few minutes. The results are not only obtained more quickly but without recourse to solvents or laborious titrimetric methods.

Near infrared spectroscopy covering the region 800–2500 nm is now being developed to determine fatty acid composition, the oil content of individual seeds, and recognition of individual vegetable oils.

6.4 Electron spin resonance (ESR) spectroscopy

ESR spectroscopy is used for the study of free radicals (odd electron species) and finds only limited use in lipid studies. It finds some application in the study of autoxidation which occurs through a free radical mechanism (section 7.2). It is also used in the study of membranes after incorporation of spin-labelled material such as the 5-doxylstearic acid shown below.

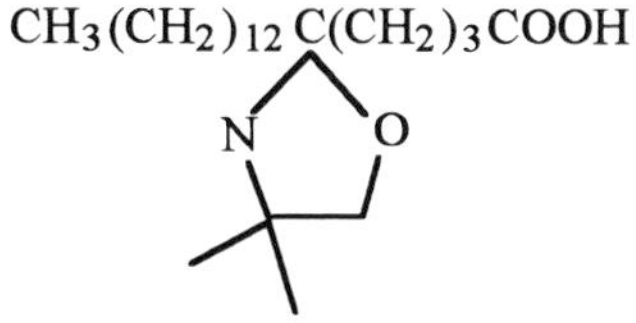

$$CH_3(CH_2)_{12}\,C(CH_2)_3COOH$$

6.5 ¹H NMR spectroscopy

¹H NMR spectroscopy is used in two ways in the study of lipids. With wide-line (low-resolution) instruments it is possible to determine the proportion of solid and liquid in a fat and the content of oil in a seed. High-resolution spectrometers, on the other hand, are used to examine solutions and give information about the solute, which may be an individual compound or a mixture, such as a natural oil or fat. Solids can be examined when the spectrometer is used in its 'magic angle' mode.

6.5.1 Low-resolution spectroscopy

Many apparently solid fats are mixtures of solid and liquid materials, the proportions of which vary with temperature. To understand the melting behaviour of such a fat it is important to know how much solid is present over a range of temperatures. Plotting solid fat content (SFC) against temperature gives a melting curve which is an important property of a fat (Figure 6.5). The steepness of the curve defines the sharpness of melting, and the melting curve also shows the temperature at which no solid remains. Solids remaining at mouth temperature give a feeling of 'waxiness' to the sample. These properties are important in confectionery and in spreading fats. The solid fat content can be measured by low-resolution pulsed ¹H NMR spectroscopy.

The percentage solids is given by the expression 100 (hydrogen nuclei in the solid phase) ÷ (all the hydrogen nuclei in the sample) and these values can be determined from observation of the relaxation signal. The signal for hydrogen nuclei in solid triacylglycerols decays quickly – less than 1% remains after 70 µs – while that from a liquid decays very slowly, requiring about 10 000 s. There are practical reasons why measurements cannot be made at the instant of the pulse but are made after 10 µs ($S_S + S_L$) and after 70 µs (S_L only). Because the first measurement can only be made after 10 µs when some of the signal from the solid phase has already decayed the S_S figure is not quite correct and has to be increased by an empirical factor f. This is determined by calibration of the system using samples of a solid plastic at known levels around 35 and 70% in liquid paraffin.

These measurements require only about 6 seconds and are used routinely for the study of margarine and other confectionery fats and of cocoa butter and similar substances.

However, though the measurement is so quick, it may have to be preceded by a lengthy tempering routine. Without controlling tempering the results would not be reproduced from day to day or between laboratories. The tempering regime varies with the kind of fat but a typical procedure for cocoa butter involves melting at 100°C then holding at

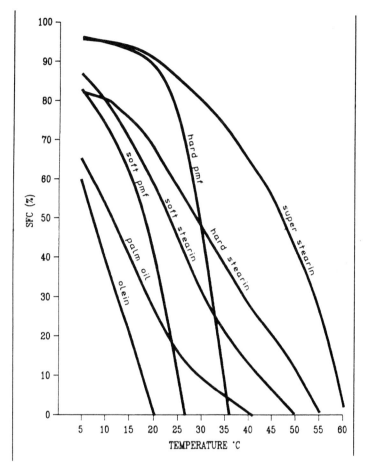

Figure 6.5 Solid fat content of some palm oil fractions by NMR (Deffense, 1995).

60°C(1 h), 0°C(1.5 h), 26°C(40 h), 0°C(1.5 h) and finally at the measuring temperature for 1 h. For many fats the long tempering period at 26°C can be omitted. If measurements are to be made at several temperatures the samples can be placed in different tubes and all can be tempered at one time in the same baths, only the final period at the measuring temperature will require different baths.

The system can be modified to measure the oil content of seeds. This information is of commercial value and can assist seed breeders and agronomists in their studies to develop improved varieties.

6.5.2 High resolution spectroscopy

A ^{1}H typical spectrum is shown in Figure 6.6. The spectrum contains several signals which can be designated in terms of chemical shift, coupling constant, splitting pattern, and area. The last of these provides useful quantitative information while the remainder give structural information. This is illustrated by discussion of some simple examples.

Methyl stearate (and other saturated esters) will show five signals. These are detailed in Table 6.9. The same signals appear in methyl oleate and methyl linoleate but oleate also contains signals for olefinic (5.35 ppm, 2H) and allylic hydrogen atoms (2.05 ppm, 4H) and for linoleate these are at 5.35 (4H) and 2.05 (4H, C8 and C14) and 2.77 ppm (2H, C11). When a double bond comes close to the methyl group as in α-linolenate and other n-3 esters the CH$_3$ signal is shifted to 0.98 ppm and oils containing such acids show two triplets at 0.98 (n-3 esters) and 0.89 ppm (n-6, n-9, and saturated esters). The areas associated with these various signals can be used to calculate semiquantitative information in terms of n-3 acids (α-linolenate), other polyene acids (linoleate), monoene acids (oleate), and other acids (saturated).

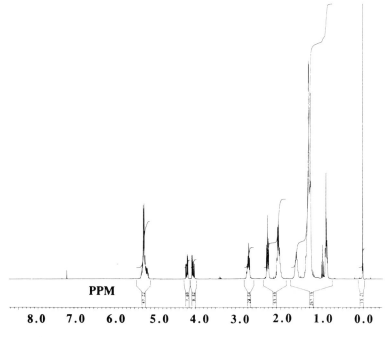

Figure 6.6 ^{1}H NMR spectrum of walnut oil. Signals at 0.9 (CH$_3$), 1.0 (CH$_3$ $_{n\text{-}3}$), 1.3 (bulk CH$_2$), 1.6 (CH$_2$-3), 2.05 (allylic), 2.3 (CH$_2$-2), 2.8 (double allylic), 4.1–4.3 (glycerol CH$_2$), 5.35 ppm (olefinic and glycerol CH).

Table 6.9 Chemical shifts (δH) for methyl alkanoates

δH (ppm)		H	
0.90	triplet	3	CH_3
1.31	(broad)	$2n$	$(CH_2)_n$
1.58	quintet	2	$-C\underline{H}_2CH_2COOCH_3$
2.30	triplet	2	$-CH_2C\underline{H}_2COOCH_3$
3.65	singlet	3	$-CH_2CH_2COOC\underline{H}_3$

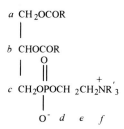

a CH_2OCOR

b $CHOCOR$

c $CH_2OPOCH_2CH_2NR'_3$

	R'	a	b	c	d	e	f
PC	CH_3	4.13⌉ 4.30⌋	5.20	3.94	4.31	3.79	3.35
PE	H	4.0–4.2	5.2	3.9	4.0–4.2	3.1	7–8.5
number of H atoms		2	1	2	2	2	9(PC) 3(PE)

Figure 6.7 ^1H NMR signals (chemical shifts, ppm) in phosphatidylcholines (PC) and phosphatidylethanolamines (PE).

Glycerol esters show further signals associated with the five hydrogen atoms in this unit. There is a one-proton signal at 5.25 ppm (CHOCOR) which overlaps with the olefinic signal and a four-proton signal split between 4.12 and 4.28 ppm (CH_2OCOR).

With phospholipids there are characteristic signals associated with phosphatidylcholines (R' = CH_3) and phosphatidylethanolamines (R' = H). These are summarized in Figure 6.7.

6.6 ^{13}C NMR spectroscopy

High-resolution ^{13}C NMR spectra are more complex than ^1H spectra and they do not provide quantitative information so easily. Nevertheless, they contain more structural information if this can be teased out of the data (chemical shifts and intensities) provided with each spectrum.

A long-chain saturated acid or ester will contain six distinct and easily recognized signals associated with the C1–3 and the ω1–3 carbon atoms. There will also be a more complex group of signals with very similar chemical shifts. These are designated as the methylene envelope and do not usually furnish useful information.

ω1   ω2   ω3   methylene envelope   C3   C2   C1

CH$_3$ CH$_2$ CH$_2$ .... CH$_2$ CH$_2$ COOH

The C1–3 signals have slightly different chemical shifts depending on the nature of the acid or ester and in the case of glycerol esters each appears as a double signal in a 2:1 ratio depending on whether the acid is associated with a primary alcohol (α chain) or a secondary alcohol (β chain) (Table 6.10). There are also some changes in the chemical shifts of short-chain acids where the effect of the CO$_2$H group at one end and the methyl group at the other interact. It is easy to distinguish C$_4$, C$_6$, and C$_8$ acids from longer-chain acids so that, for example, useful information can be obtained about milk fats and milk products.

Glycerol esters also have signals for the glycerol carbon atoms. These will be discussed later along with different kinds of glycerol esters.

If one or more unsaturated centres or some other functional group (hydroxy, epoxy) is present in the long chain it will generate its own set of signals. For example, in the simple case of one double bond there will

Table 6.10 Chemical shifts (ppm) of C1–3 signals from glycerol tripalmitate

TAG chain	C1	C2	C3
α-chain	173.27	34.07	24.89
β-chain	172.86	34.24	24.94
difference	0.41	0.17	0.05

Table 6.11 Chemical shifts (ppm) for olefinic carbon atoms in isomeric 18.1 acids

Δ	cis			trans		
8	130.17	129.65	(0.52)[a]	130.67	130.13	(0.54)[a]
9	130.09	129.78	(0.31)	130.54	130.23	(0.31)
10	130.00	129.83	(0.17)	130.47	130.31	(0.16)
11	129.96	129.89	(0.07)	130.43	130.34	(0.09)
12		129.94			130.41	

[a]Difference in chemical shifts of the two olefinic carbon atoms. The higher chemical shift relates to the carbon further from the carboxyl group, e.g. for oleic acid C10(130.09) and C9(129.78).

Table 6.12 Chemical shifts (ppm) for olefinic signals in oleic, linoleic, and α-linolenic glycerol esters

Carbon atom	Oleate	Linoleate	Linolenate
9	129.70	[129.98]	[130.19]
10	[129.98]	128.09	127.77
12	–	127.91	128.24
13	–	[130.19]	128.29
15	–	–	127.13
16	–	–	131.93

Overlapping signals are shown in brackets.

Table 6.13 Chemical shifts (ppm) of some allylic carbon atoms

Acid/ester configuration		C8	C11	C14
18:1 (9)	c	27.19	27.24	–
	t	32.58	32.63	–
18:2 (9,12)	c,c	27.23	25.70	27.23
	c,t	27.18	30.55	32.64
	t,c	32.67	30.56	27.22
	t,t	32.59	35.68	32.59
18:3 (9,12,15)	c,c,c	27.24	25.67	25.57

usually be two signals for the olefinic carbon atoms and signals for the carbon atoms α (allylic) and γ to the double bond. The olefinic chemical shifts differ between *cis-* and *trans*-isomers and with the position of the double bond in the chain. When the monoene ester is a pure compound or the major component in a mixture it is possible to determine the double bond position. Some typical chemical shifts useful in the examination of partially hydrogenated oils are listed in Table 6.11. Data on oleate, linoleate and α-linolenate are collected in Table 6.12 and information about allylic signals in Table 6.13.

The influence of a double bond (or other functional group) is not confined to the allylic carbon atom where it is most marked and different for *cis-* and *trans*-isomers (Table 6.13). It may affect chemical shifts up to six carbon centres away and this may become apparent on the C1–3 and ω1–3 signals. With the mixture of fatty acids present in an oil or fat the ω1–3 signals no longer appear as single signals but as clusters of two or more signals showing, for example, the presence of *n*-3, *n*-6 and *n*-9 and saturated acyl chains (Table 6.14). In a vegetable oil the *n*-3 and *n*-6 signals are most likely to be related to α-linolenate and linoleate, respectively.

The effect of the Δ9 double bond in oleate, linoleate, and α-linolenate on the C1–3 signals is quite small but can be observed if the spectrum is gathered over a longer time than usual. Chemical shifts are changed more

Table 6.14 Chemical shifts (ppm) for ω1–3 signals in saturated and unsaturated glycerol esters

Ester	ω1	ω2	ω3
sat			} 31.95
n-9	} 14.13	} 22.60	
n-7			31.83
n-6	14.09	22.71	31.55
n-3	14.29	20.57[a]	[b]

[a]This is also an allylic carbon atom.
[b]An olefinic carbon atom.

Table 6.15 Chemical shifts (ppm) for glycerol esters of petroselinic and other saturated and unsaturated acids in carrot seed oil: content and distribution of petroselinic acid

	C1		C2		C3	
	α	β	α	β	α	β
petroselinic	173.05	172.66	33.95	34.11	24.51	24.54
other	173.16	172.76	34.05	34.21	24.87	24.90
petroselinic acid[a]						
total (%)	70.9		70.3		70.5	
α chain	83		81			
β chain	48		51			

[a]Based on intensity values not listed here.

Table 6.16 Chemical shifts (ppm) of glycerol carbon atoms in mono-, di-, and triacylglycerols

	CDCl₃		CDCl₃-CD₃OD(2:1)	
Carbon atoms	β	α	β	α
glycerol ester				
1-mono	70.27	65.04, 63.47	70.24	65.58, 63.46
2-mono	74.97	62.05	75.36	61.06
1,2-di	72.25	62.20, 61.58	72.40	62.92, 60.82
1,3-di	68.23	65.04	67.71	65.36
tri-	68.93	62.12	69.41	62.50

if the double bond is closer to the acyl group as in oils containing DHA ($\Delta4,7,10,13,16,19$), EPA ($\Delta5,8,11,14,17$), AA ($\Delta5,8,11,14$), GLA ($\Delta6,9,12$), or petroselinic acid ($\Delta6$). Effects can be observed in signals both for the α and β chains so it is possible to estimate the levels of such acids in each chain in natural glycerol esters. An example for the petroselinic acid in carrot seed oil is given in Table 6.15.

In a similar way it is possible to recognize and locate functional groups such as hydroxy, epoxy, acetylenic, cyclopropene, and branched chains.

The characteristic signals related to the polar head groups of mono-, di-, and triacylglycerols (Table 6.16) and in phospholipids (Table 6.17) have been identified and can be used for semiquantitative analysis.

^{13}C NMR spectroscopy has been applied to epoxidized oils, and to partially hydrogenated fats.

There is enough information in this account to interpret the spectrum of safflower oil given in Figure 6.8.

6.7 Mass spectrometry

When a molecule (M) is bombarded in the vapour phase at low pressure with electrons of sufficient energy the molecule becomes ionized to give a molecular ion M^+. At greater electron energies the ionized molecule decomposes, usually in several different ways, to give one charged (B^+) and one uncharged particle (A).

$$M \xrightarrow[-2e^-]{+e^-} M^+ \longrightarrow A + B^+$$

A mass spectrometer is a device for producing and examining positively charged particles which are separated according to their mass to charge ratio (m/z, z is usually one). With high-resolution instruments this value can be measured with such accuracy as to indicate the molecular formula of each ion.

In the electron-impact mass spectrum (EIMS) of methyl stearate there is a molecular ion peak at $m/z = 298$, a series of prominent peaks caused by oxygen-containing fragments of the type $(CH_3OCO(CH_2)_n]^+$ such as 87, 101, ... 241, a peak at M-31 (OMe) characteristic of methyl esters, and the base peak (largest individual peak) at $m/z = 74$ resulting from cleavage between C2 and C3 to give the fragment $[CH_3OC(OH) = CH_2]^+$.

Table 6.17 Chemical shifts (ppm) for signals in the polar head groups of triacylglycerols, phosphatidylethanolamines and phosphatidylcholines derived from soybean oil

Source	TAG	PE	PC
C1α	173.91	174.11	174.14
C1β	173.49	173.74	173.79
G1	62.46	62.86	62.86
G2	69.38	70.72, 70.62[a]	70.67, 70.62[a]
G3	62.46	64.06, 63.99[a]	63.88
CH$_2$O	–	61.91, 61.88[a]	59.44, 59.38[a]
CH$_2$N$^+$	–	40.87	66.65
Me$_3$N$^+$	–	–	54.30

[a]Signals split by phosphorus.
G = glycerol carbon atoms.

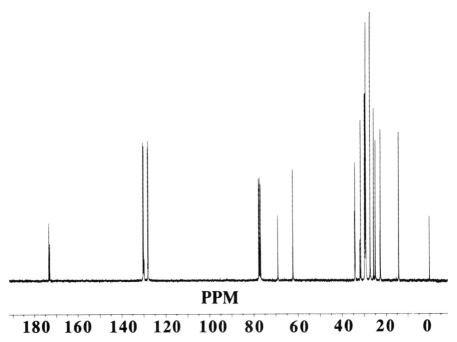

Figure 6.8 Chemical shifts (ppm) of ^{13}C NMR spectrum of safflower oil. C1, 173.21 (α), 172.80 (β); C2, 34.19 (β), 34.03 (α); C3, 24.85; ω1, 14.09, ω2, 22.71 and 22.60; ω3, 31.93 and 31.54; glycerol; 68.89 (β) and 62.10 (α) olefinic, 130.20, 129.98, 129.70, 128.07, 127.89, (Table 6.12); allylic 27.21 (*cis*) and 25.64 (L11); methylene envelope 29.11–29.78 ppm (11 signals).

When electron-impact mass spectroscopy is applied to an unsaturated ester such as methyl oleate the results are not very encouraging. Ions corresponding to $[M]^+$, $[M–31]^+$, and the fragment of $m/z = 74$ are present but fragment ions which might have been used to indicate double bond position are of no value. Under electron bombardment the double bond is labile and all 18:1 esters give virtually the same mass spectrum. This difficulty has been overcome in two ways.

First, attempts were made to 'fix' the double bond by conversion to some chemical derivative. This is chosen to give a mass spectrum with significant fragment ions which indicate the position of the double bond. There are several ways of doing this which work well with monoenes but are generally less satisfactory with dienes and more-unsaturated esters. The problem of interpretation increases with the number of double bonds.

As an alternative to this procedure the methyl esters are replaced by amides or nitrogen-containing esters which give more useful spectra. These derivatives must meet two criteria: they must give useful spectroscopic information but they should also have good chromatographic properties so that mixed products can be separated from each other before they are examined spectroscopically in a GC-MS system.

Glycerol esters and phospholipids can be examined by tandem mass spectrometry.

6.7.1 Derivatives of olefinic compounds

The identification and location of oxygen-containing functional groups (epoxy, hydroxy, oxo) presents no difficulty and fragmentation patterns are indicated in Figure 6.9. Olefinic compounds can be converted to oxygenated compounds *via* epoxides, diols, and mercuriacetates as detailed in Figure 6.9. As already indicated they give complex spectra which may be difficult to interpret as the number of olefinic centres increases. This is particularly true when the product from a monoene is a mixture of two compounds and the product from a diene contains four compounds and so on. Useful derivatives are produced from epoxides (two hydroxy compounds, two hydroxy methoxy compounds, two hydroxy dimethylamino compounds, or two ketones), from diols (bismethyl ether, bis (trimethylsilyl) ether, acetonoate, and methyl or butyl boronate), or from the mercuriacetates (two bromo methoxy compounds or two methoxy compounds). Structures, reagents, and fragmentation patterns are summarized in Figure 6.9. For polyunsaturated compounds it is best to prepare the polyols and from these the poly(trimethylsilyl) ethers.

Another derivative successfully used with monoenes is the bis(methylthio) compound prepared by reaction with dimethyldisulphide and iodine:

$$-CH=CH- \xrightarrow[\text{I}_2]{\text{Me}_2\text{S}_2} -CH(SMe)CH(SMe)-$$

With dienes reaction appears to be restricted to one double bond so that linoleate gives two products which can be identified by their fragment ions:

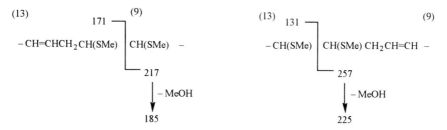

A C_{21} acid with five double bonds was shown to be the *n*-3 isomer (Δ6,9,12,15,18) by reduction (N_2H_4, H_2O_2) to a partially reduced product, isolation of the monoenes, and MS examination of their bis(methylthio) derivatives. Fragments at 257, 175, 143 (Δ6), 215, 217, 185, (Δ9), 173, 259, 227, (Δ12), 131, 301, 269 (Δ15), and 89, 343, 311 (Δ18)

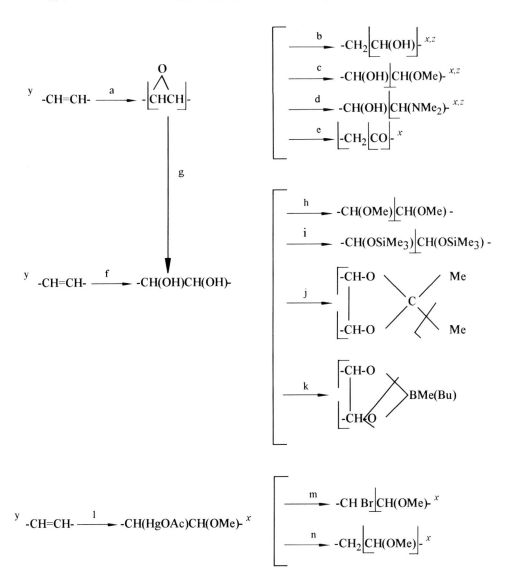

Figure 6.9 Derivatives of olefinic esters used for mass spectrometric location of double bond positions in the alkenoate $CH_3(CH_2)_n$ $CH=CH(CH_2)_m$ CO_2Me. x; these compounds are formed as mixtures of two regioisomers but only one structure is shown. y; end methyl group to the left, carboxyl group to the right. z; compounds with a free hydroxyl group sometimes give better results as the trimethylsilyl ester. These symbols indicate position of fragmentation with the charge on the fragment indicated by the lower horizontal line. Where more than one such symbol is shown these are alternative fragmentations. *Reagents: a, peracid; b, LiAlH₄; c, BF₃, MeOH; d, Me₂NH; e, NaI; f, OsO₄; g, AcOH followed by hydrolysis; h, MeI; i, (Me₃Si)₂NH; j, COMe₂; k, MeB(OH)₂ or BuB(OH)₂; l, MeOH, Hg(OAc)₂; m, bromine (or iodine); n, NaBH₄.

corresponding to those shown above for linoleate, are sufficient to fix the position of the five double bonds.

6.7.2 Nitrogen-containing esters and amides

Picolinyl esters are now widely used for MS studies of long-chain acids. They are made by the following two-step process:

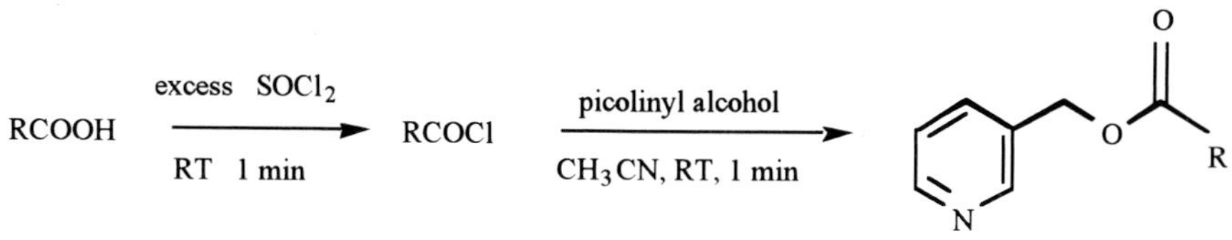

These esters show good gas chromatographic separation at about 50°C above that used for the corresponding methyl esters. Spectra obtained at 25 eV are preferred to those obtained at the more usual 70 eV. The spectra show a molecular ion $[M]^+$ which itself provides useful information, along with common peaks at $m/z = 93, 108, 151,$ and 164 (Figure 6.10), and a further series of fragment ions with even mass. These are the major signal in each cluster and are usually sufficient to provide information about branching and about double bond position. Data for stearate, oleate, linoleate, and α-linolenate are summarized in Table 6.18. The following is a typical fragmentation for stearate occurring between C_{10} and C_{11} and producing the C_{10} ester fragment:

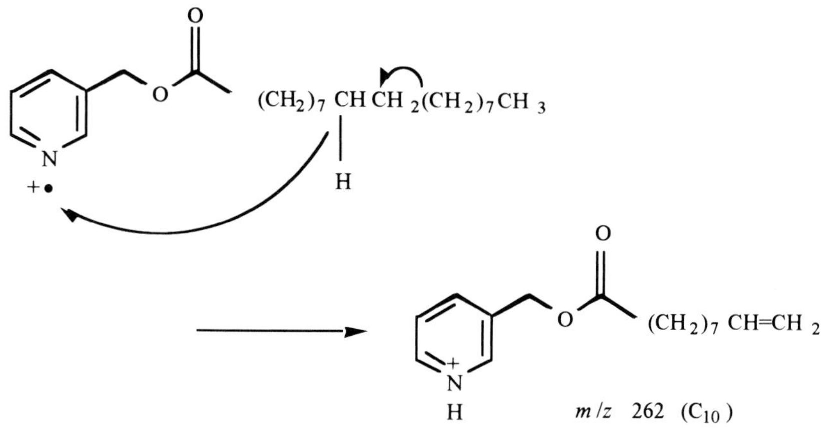

Comparison of the fragment ions for stearate and oleate (Table 6.18) show how it is possible to determine the double bond position. The features to note are: (a) the C_{10} to C_{17} fragments are two mass units lower for oleate than for stearate while those up to C_8 have the same mass;

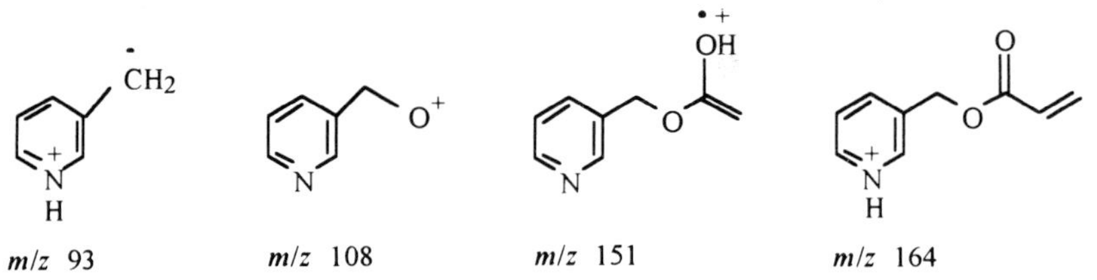

m/z 93   *m/z* 108   *m/z* 151   *m/z* 164

Figure 6.10 Fragments common to all picolinyl esters.

Table 6.18 Significant fragment ions in the mass spectra of the picolinyl esters of stearic, oleic, linoleic and α-linolenic acids

Ester	$[M]^+$	17	16	15	14	13	12	11	10	9	8	7	6	5	4
						Number of carbon atoms in the acyl chain of the fragment									
18:0	375	360	346	332	318	304	290	276	262	248	234	220	206	192	178
18:1	373[a]	358	344	330	316	302	288[a]	274[a]	260	†	234	220	206	192	178
18:2	371[a]	356	342	328	314	300	†	274	260	†	234	220	206	192	178
18:3	369[a]	354	340	†	314	300	†	274	260	†	234	220	206	192	178

[a]Enhanced signal.
†Missing fragment, gap of 26 mass units between adjacent signals.
    Gap of 40 units between somewhat enhanced signals.

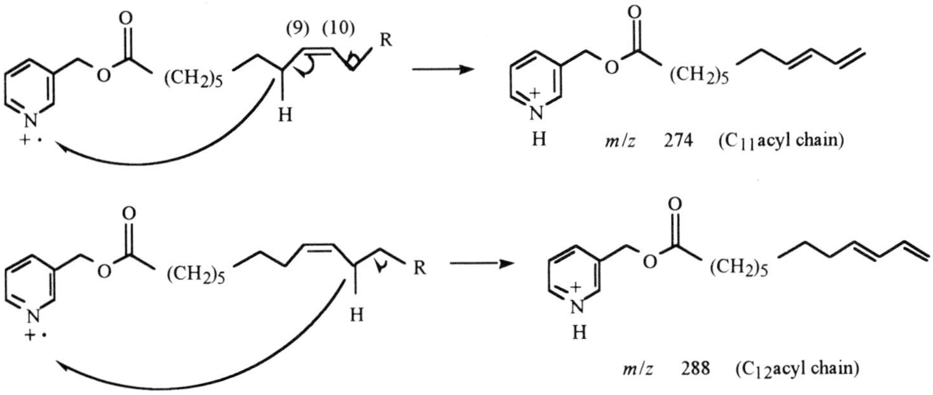

Figure 6.11 Fragments produced from picolinyl oleate involving the allylic groups.

(b) there are two enhanced signals for the C_{11} and C_{12} fragment ions which are related to allylic groups at C_8 and C_{11} (Figure 6.11); and most significantly (c) the C_9 fragment is missing and the C_8 and C_{10} fragments differ by only 26 mass units.

These concepts are extended to linoleate and α-linolenate. Missing fragments (C_9 and C_{12} for linoleate and C_9, C_{12} and C_{15} for α-linolenate) lie between adjacent fragment ions differing by 26 mass units. Sometimes it is easier to recognize slightly enhanced fragments differing by 40 mass units (Table 6.18). The gaps of 26 and 40 mass units are related to the olefinic centres and define their position. With yet more double bonds the interpretation becomes more complex but it is usually possible to recognize the unsaturated centre closest to the methyl group. It may then be assumed that the remaining unsaturated centres assume the usual methylene-interrupted pattern.

Branched methyl groups are located by observation of a gap of 28 mass units in the usual progression of numbers differing by 14. For example with 16-methyloctadecanoic picolinyl ester (anteiso) there is no M-43 peak because of the loss of a CH_3CH unit rather than the usual CH_2.

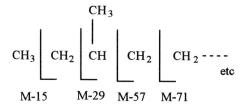

Other derivatives of a similar nature which have been used include N-acylpyrrolidides, 2-alkylbenzoxazoles, and 2-alkyl-4,4-dimethyloxazoles (DMOX). Long-chain alcohols can be examined as their nicotinic acid esters. These molecules are formulated in Figure 6.12 along with the methods for their preparation.

Mass spectrometry is also used in conjunction with compounds containing stable isotopes (^2H, ^{13}C) and to distinguish between natural and synthetic fragrances.

6.7.3 Tandem mass spectrometry

Tandem mass spectrometry (also known as MS/MS) involves a study of the further fragmentation of fragment ions achieved in the presence of a collision gas through activation. This is described as collision-induced dissociation (CID) or collision-activated dissociation (CAD).

$$m_1^+ \longrightarrow m_2^+ + m_0$$

| parent | daughter | neutral |
| ion | ion | fragment |

An example of this technique is the study of castor oil as its trimethylsilyl ether (TMS) derivative (Hogge et al., 1991). Partial fractionation (HPLC) followed by MS/MS leads to the recognition of 11 different glycerol esters. These contain two ricinoleic acid chains and a third acyl chain which may

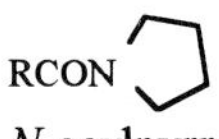

RCON⟨pyrrolidine⟩
N-acylpyrrolidide

reaction of ester with excess of pyrrolidine
and acetic acid at 100°C for 30 min

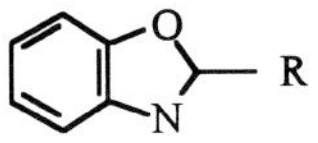

2-alkylbenzoxazole

reaction of acid with 2-aminophenol and
polyphosphoric acid at 70°C for 30–60 min

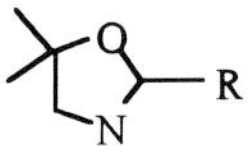

2-alkyl-4,4-dimethyl-
oxazole

reaction of acid with $Me_2C(OH)CH_2NH_2$ at
180°C for 2 h in a nitrogen atmosphere

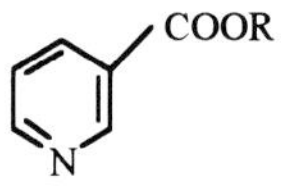

COOR

nicotinic ester
of long-chain alcohol

reaction of alcohol with nicotinyl chloride
hydrochloride in acetonitrile

Figure 6.12 N-containing derivatives of long-chain acids and alcohols used for mass spectrometry.

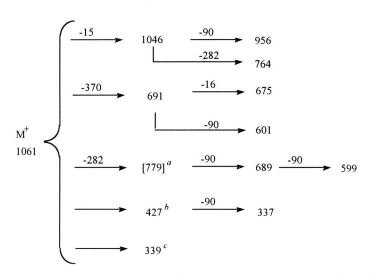

Figure 6.13 Fragments obtained by MS-MS from glycerol esters in castor oil containing ricinoleic (2 moles) and oleic acid (1 mole). (Hogge *et al.* (1991)). *a*; this fragment was not observed. *b*; RCO + 74 (ricinoleic TMS ether). *c*; RCO + 74 (oleic). 90, 282, and 370 represent OTMSH, oleic acid, and ricinoleic (OTMS) acid respectively.

be ricinoleic acid, another hydroxy acid, or a non-hydroxy acid. Several of the daughter ions indicate the nature of these acyl groups including $[RCO+74]^+$ and fragments in which RCOOH is missing. As an example, the fragments detailed in Figure 6.13 came from a glycerol ester containing oleic acid along with the two ricinoleic chains.

A second example (Currie and Kallio, 1993) is concerned with the triacylglycerols of human milk fat. Ammonia negative ion chemical ionization mass spectrometry gives 29 $[M\text{-}H]^-$ ions. These range from 38–54 carbon atoms, molecular weight 666.7–886.7, and have 0–4 double bonds. Twenty-eight different fatty acid molecular weights are identified. The composition of each $[M\text{-}H]^-$ cluster is determined from a study of the $[RCO_2]^-$ daughter ions and the distribution of acyl groups between the sn-1/3 and sn-2 positions from a study of the $[M\text{-}H\text{-}RCO_2H\text{-}100]^-$ fragments. The most abundant glycerol ester is the symmetrical 1,3-dioleate 2-palmitate which makes up about 10% of the total.

Bibliography

Christie, W.W. (ed.) (1992) *Advances in Lipid Methodology – One*, The Oily Press, Dundee.
Christie, W.W. (ed.) (1993) *Advances in Lipid Methodology – Two*, The Oily Press, Dundee.
Currie, G.T. and M. Kallio (1993) Triacylglycerols of human milk: rapid analysis by ammonia negative ion tandem mass spectrometry, *Lipids*, **28**, 217–222.
Deffense, E. (1995) *Lipid Technol*, **7**, 34.
Evershed, R.P. (1994) Application of modern mass spectrometric techniques to the analysis of lipids, in *Developments in the Analysis of Lipids* (eds J.M.P. Tyman and M.H. Gordon) Royal Society of Chemistry, Cambridge.
Grati, N. and K. Sato (eds) (1988) *Crystallisation and Polymorphism of Fats and Fatty Acids*, Dekker, New York.
Gunstone, F.D. (1993) High resolution ^{13}C nmr spectroscopy of lipids, in *Advances in Lipid Methodology – Two* (ed. W.W. Christie). The Oily Press, Dundee, pp. 1–68.
Gunstone, F.D. and V.K.S. Shukla (1995) NMR of lipids, in *Annual Reports on NMR Spectroscopy*, Vol. 31 (eds Webb, Belton, and McCarthy) Academic Press, London, pp. 219–237.
Gunstone, F.D. (1994) ^{13}C NMR of lipids, in *Developments in the Analysis of Lipids* (eds J.M.P. Tyman and M.H. Gordon) Royal Society of Chemistry, Cambridge.
Hachiya, I. *et al.* (1990) Fat polymorphism and chocolate crystallisation, *Lipid Technol.*, **2**, 34–37.
Hagemann, J.W., W.H. Tallent, J.A. Barve, I.A. Ismail and F.D. Gunstone (1975) *J. Amer. Oil Chem. Soc.*, **52**, 204–207.
Harvey, D.J. (1992) Mass spectrometry of picolinyl and other nitrogen-containing derivatives of lipids, in *Advances in Lipid Methodology – One* (ed. W.W. Christie). The Oily Press, Dundee, pp. 19–80.
Hogge, L.R. *et al.* (1991) Characterisation of castor bean neutral lipids by mass spectrometry/mass spectrometry, *J. Amer. Oil Chem. Soc.*, **68**, 863-868.
Lange, H. and M.J. Schwuger (1981) Physico-chemical properties of fatty alcohols, in *Fatty Alcohols* (ed. K. Henkel), Henkel, Dusseldorf, pp. 87–119.
Larsson, K. *et al.* (1994) Physical properties, in *The Lipid Handbook* (eds F.D. Gunstone, J.L. Harwood, and F.B. Padley) Chapman and Hall, London, pp. 401–560.
Larsson, K. (1994) *Lipids – Molecular Organisation, Physical Functions, and Technical Applications*, The Oily Press, Dundee.

Povcy, M.J.W. Analysis of lipid structure by neutron diffraction, in *Developments in the Analysis of Lipids* (eds J.M.P. Tyman and M.H. Gordon) Royal Society of Chemistry, Cambridge.

Pryde, E.M. (ed.) (1979) *Fatty Acids*, AOCS, Champaign.

Sadeghi-Jorabchi, M. (1994) Fourier-transform infrared spectroscopy in lipid analysis, *Lipid Technol.*, **6**, 146–149.

Waddington, D. (1986) Application of wide-line nuclear magnetic resonance in the oils and fats industry, in *Analysis of Oils and Fats* (eds R.J. Hamilton and J.B. Rossell) Elsevier, London, pp. 341–400.

7 Reactions associated with double bonds

7.1 Catalytic hydrogenation, chemical reduction, biohydrogenation

Partial reduction of unsaturated oils by heterogeneous catalytic reduction is an important way of modifying such materials. Each year millions of tons of soybean oil and other vegetable oils containing oleic, linoleic, and linolenic acid are reduced. Fish oils with polyunsaturated C_{18}, C_{20}, and C_{22} acids are subject to similar processing.

The need for this change and details of the process have been described in section 4.3. This section is concerned with a fuller report of the chemical changes occurring during partial hydrogenation and with some account of other reduction processes. These changes relate to olefin reduction: reaction of the ester function to produce alcohols is covered in section 8.6.

7.1.1 Catalytic hydrogenation

During partial hydrogenation some double bonds are saturated while others undergo stereomutation (conversion of *cis*- to *trans*-isomers) and/or double bond migration. Although the industrial process is a reaction of triacylglycerols understanding of this reaction came first from the study of individual methyl esters.

(a) *Methyl oleate.* Complete hydrogenation of methyl oleate gives only methyl stearate but if the reaction is stopped before completion then, in addition to oleate and stearate, several octadecenoate isomers (iso-oleate) are present. These are a mixture of the *cis* and *trans* forms of several positional isomers. The mechanism of this change is discussed later:

$$\text{etc.} \quad \Delta 11 \; \rightleftharpoons \; \Delta 10 \; \rightleftharpoons \; \boxed{\Delta 9} \; \rightleftharpoons \; \Delta 8 \; \rightleftharpoons \; \Delta 7 \quad \text{etc.}$$

(b) *Methyl linoleate.* Reduction of methyl linoleate is 5–100 times quicker than reduction of oleate depending on the reaction conditions (this relates to k_2, k_3, and S_1, see section 4.3). The changes occurring are set out in Figure 7.1. If the first reaction is hydrogenation then linoleate will give a mixture of $\Delta 9$ and $\Delta 12$ *cis*-monoenes each of which will then react independently. If double bond migration occurs before reduction then the reaction mixture will contain three kinds of dienes: (i) conjugated (mainly

Δ9,11 and Δ10,12); (ii) methylene-interrupted dienes (unreacted lino-leate); and (iii) non-methylene-interrupted dienes (mainly Δ8,12- and Δ9,13-, etc.). The rate of reaction of these diene types is given by conjugated > methylene-interrupted >> non-methylene-interrupted. Conjugated dienes have been observed in partially hydrogenated products only at very low levels because they are quickly reduced to monoenes.

A recent study of the 18:2 isomers remaining (2.1%) in a partially hydrogenated canola oil showed that the major components were the 9c13t (0.58%) 9c12t (0.47%), 9t12c (0.33%), and 9c15c (0.27%) isomers and others at lower levels.

Some catalysts, such as copper chromite, reduce methylene-interrupted dienes and conjugated dienes only. There is no reduction of monoenes or of non-methylene-interrupted dienes.

(c) *Methyl linolenate.*   Partial reduction of methyl linolenate with three double bonds becomes quite complicated. When double bond migration produces conjugated systems then rapid reduction of these will follow. Methylene-interrupted dienes will be moderately reactive but non-methylene-interrupted dienes are less reactive and may remain in the partially hydrogenated product. Examples of these esters are listed in Table 7.1.

One study has shown that in the reduction of linoleate 95–98% is converted first to monoene and only a small portion (2–5%) is hydrogenated directly to stearate. With linolenate about 79% is reduced to diene, up to 21% is reduced directly to monoene, while direct reduction to stearate is only observed at high pressure and then at levels not exceeding 1%.

(d) *Reaction mechanism.*   The competing processes of reduction, stereo-mutation, and double bond migration are usually discussed in terms of a reversible reaction occurring through a half-hydrogenated intermediate as first suggested by Horiuti and Polanyi (Figure 7.2).

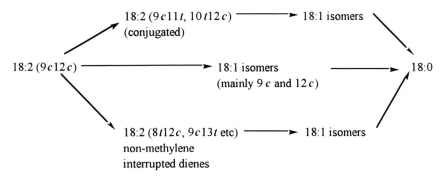

Figure 7.1 Hydrogenation of methyl linoleate.

Table 7.1 C_{18} trienes and dienes formed during the partial hydrogenation of methyl linolenate[a]

Pattern of unsaturation	Trienes		Dienes	
methylene-interrupted	9,12,15	9,12	12,15	
conjugated	9,11,15	9,11	10,12	
	10,12,15			
	9,12,14			
	9,13,15			
non-methylene-interrupted	not defined	9,15	9,13	9,14
		10,15	10,13	

[a]These compounds may be present in *cis* and *trans* forms. A sample of partially hydrogenated soybean oil contained 3.9% of trienes. These were mainly unreacted linolenate (2.7%) but the 9t12t15c, 9t12c15c, and 9c12c15t isomers were also present.

1. Hydrogen and olefinic ester are adsorbed at active sites on the surface of the metallic catalyst.
2. Reaction between olefin and hydrogen gives a half-hydrogenated species linked to the catalyst through the other olefinic carbon atom.
3. If this species reacts with a second hydrogen atom then the olefin has been reduced and desorbs from the catalyst surface.
4. If, instead, reaction (ii) is reversed then the half-hydrogenated species will lose any one of four hydrogen atoms (see Figure 7.2). Loss of one of these will produce the original olefin in its *cis* form, loss of another will give the *trans*-isomer of the original olefin, and loss of the remaining two will give the *cis*- and *trans*-isomers of compounds in which the double bond has shifted.

7.1.2 Other chemical reductions

Complete hydrogenation of an unsaturated ester gives the saturated or perhydro ester. This is easily effected in the laboratory with palladium on charcoal and may be a useful step in the identification of a fatty acid of unknown structure (section 1.14.5).

Partial reduction of acetylenic acids gives olefinic acid with *cis* (Lindlar's catalyst) or *trans* configuration (sodium in liquid ammonia). These are important steps in the synthetic routes proceeding through acetylenic intermediates (section 1.16.2).

Non-catalytic reduction is effected by hydrazine in the presence of oxygen or some other oxidizing agent or with hydroxylamine and ethyl acetate in a nitrogen atmosphere. In both cases the reactive species is di-imine (N_2H_2) and reduction involves a *cis* addition of hydrogen with no migration of unreacted double bonds. Using N_2D_4 in place of N_2H_4 it is possible to convert *cis*-alkenes to *erythro*-dideuterio compounds and *trans*-alkenes to *threo*-dideuterio compounds. Partial reduction with di-imine has

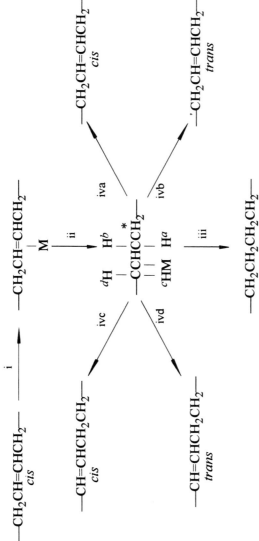

Figure 7.2 The hydrogenation and other reactions of an olefinic centre occurring through the half-hydrogenated intermediate. (i) adsorption of alkene on metal catalyst. (ii) reaction with hydrogen atom on metal catalyst. (iii) desorption from metal catalyst. (iv a–d) reversal of step (ii) with hydrogen atom *a–d* being removed to give the isomeric alkenes shown.

* This half-hydrogenated product is linked to a second hydrogen atom at one end of the olefinic system and to metal at the other. The process could occur the other way round to give a half-hydrogenated product which would furnish the alternative products of double bond migration.

been employed as a step in the identification of double bond position in polyenes (see section 1.14.5). Partial reduction of γ-linolenic acid gives all possible reaction products but not in equal yield (Table 7.2). For C_{18} compounds the relative rates of reaction are Δ3 (1.57) > Δ15 (1.28) > Δ12 (1.04) > Δ9 (1.00) > Δ6 (0.91).

7.1.3 Biohydrogenation

Although ruminant animals (cows, sheep) ingest polyene acids from seed meals and from grass their depot and milk fats are rich in saturated and monoene acids and contain only low levels of polyene acids. The monoene acids include stereoisomers and positional isomers. For example, cow milk fat contains Δ6–Δ16 18:1 acids. This situation arises from the bio-hydrogenation of about 90% of the polyene acids in the rumen through the activity of rumen bacteria. Under the influence of bacteria free acids are first liberated from the lipids (triacylglycerol, glycolipid, phospholipid) and then subjected to double bond migration and to partial or complete hydrogenation. Two examples are cited here: *Butyrivibrio fibrosolvens* converts linoleic acid to *trans*-vaccenic acid by the route shown while bovine rumen liquor has been shown to reduce linoleate and α-linolenate as indicated:

18:1 (9c12c)   ⟶   18:2 (9c11t)   ⟶   18:1 (11t)
18:2 (9c12c)   ⟶   18:0
18:3 (9c12c15c)   ⟶   18:1 (15c)

It is worthy of note that the major dietary sources of *trans*-acids come from the partial catalytic hydrogenation of vegetable fats in the form of spreading fats and cooking fats or from biohydrogenation of similar materials in the form of dairy and meat products of ruminant origin.

Because of the concern expressed in some quarters about the possible risks associated with saturated dairy fats a special feeding regimen has been developed for ruminants in which dietary unsaturated lipids are protected

Table 7.2 C_{18} acids formed during partial reduction of γ-linolenic acid with hydrazine

	18:0	18:1			18:2			18:3
Double bonds		6	9	12	6,9	6,12	9,12	
(A)	1.4	1.8	2.5	2.1	7.5	11.9	10.1	62.7
(B)	3.8	4.8	6.7	5.6	20.1	31.9	27.1	–
			17.1			79.1		

(A) Observed yield.
(B) Yield calculated for reduced products only.

from bacteria during their passage through the rumen by encapsulation in formaldehyde-treated protein. As a result, both the milk fat and the depot fat are more unsaturated. This change in fatty acid profile will affect oxidative stability and may influence the flavour developed during cooking.

7.2 Autoxidation and photo-oxygenation

7.2.1 Introduction

Lipid oxidation is an important reaction occurring between unsaturated lipids and atmospheric oxygen. The oxidation process is accelerated by metals, light, heat, and by several initiators. It can be inhibited by antioxidants. Lipid oxidation occurs under enzymic (section 7.4) and non-enzymic conditions and the latter can operate through autoxidation or photo-oxidation. The reactions occur naturally with a wide variety of substrates, under a range of conditions, and are complex. The basic features of the reactions, as presently understood, are set out here but many details remain to be explained and these important processes continue to attract a good deal of investigation of the changes occurring both *in vitro* and *in vivo*.

The primary oxidation products are allylic hydroperoxides. The double bonds remain but may have changed position and/or configuration from their original form. The hydroperoxides may undergo further changes as set out in Figure 7.3.

The oxidation of monoene and polyene acids differ slightly from each other while autoxidation and photo-oxidation give similar, but not identical, products.

Recent developments have resulted from improved methods of separating products (especially chromatography) and of identifying them (especially spectroscopy).

7.2.2 Autoxidation

Autoxidation is a radical chain process. This indicates that the inter-mediates are radicals (odd electron species) and that reaction involves an initiation step and a propagation sequence which continues until the operation of one or more termination steps (Figure 7.4). There is usually an induction period during which oxidation occurs only slowly, followed by a period of more rapid reaction.

The detailed nature of the initiation step is not fully understood but three reactions may be involved: (i) the metal-catalysed decomposition of hydroperoxide produces initiating radicals (it is very difficult to obtain

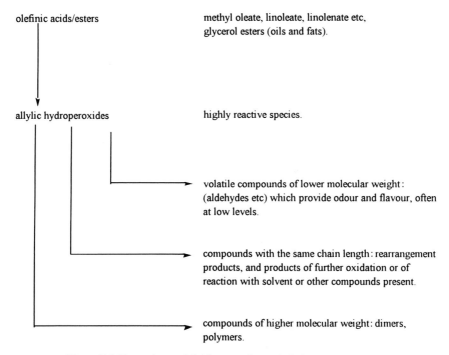

Figure 7.3 Formation and further reactions of allylic hydroperoxides.

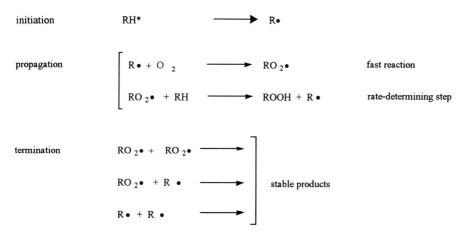

Figure 7.4 Olefin autoxidation. * RH represents an olefinic compound in which H is attached to an allylic carbon atom.

olefinic compounds entirely free of oxidation products); (ii) photo-oxygenation (a very rapid reaction) may be responsible for the first-formed hydroperoxides; and (iii) thermal initiation is possible in a heated sample. Radical sources may be added to promote the reaction when this is desired. The structure of the radical R• depends on the fatty acid being oxidized and influences both the composition of the product and the rate of reaction.

In the propagation sequence, given an adequate supply of oxygen, the reaction between alkyl radical (R•) and molecular oxygen is fast and the reaction of peroxy radical (ROO•) with another olefinic molecule is rate-determining.

Attempts to inhibit the autoxidation process are based on hindering the induction step and on promoting the termination reaction so that the propagation sequence proceeds through as few cycles as possible. The nature and role of antioxidants is discussed in section 7.3. A more detailed description of reaction products follows in section 7.2.5.

7.2.3   Photo-oxygenation

Photo-oxygenation involves mainly interaction between a double bond and highly reactive singlet oxygen produced from ordinary triplet oxygen by light in the presence of a sensitizer such as chlorophyll, erythrosine, rose bengal, or methylene blue.

$$\text{sens} + h_\nu \longrightarrow {}^1\text{sens} \xrightarrow{{}^3O_2} \text{sens} + {}^1O_2$$

Activated sensitizer can also interact with fatty acid to produce an alkyl radical which will react in the same way as in autoxidation (type-1 photosensitized oxidation) but activation of the oxygen (type-2 photosensitized oxidation) is generally more important.

This type-2 photo-oxidation differs from autoxidation in several import-ant respects: (i) it involves reaction with singlet oxygen produced from triplet oxygen by light and a sensitizer; (ii) it is an ene reaction and not a radical chain process; (iii) it displays no induction period; (iv) it is unaffected by the antioxidants normally used to inhibit autoxidation but is inhibited by singlet oxygen quenchers such as carotene; (v) it is confined to olefinic carbon atoms but is accompanied by double bond migration with

Table 7.3 Relative rates of autoxidation and photo-oxygenation of oleate, linoleate, and linolenate

Process	18:1	18:2	18:3
autoxidation (i)	1	27	77
photo-oxygenation (ii)	3×10^4	4×10^4	7×10^4
(ii) ÷ (i)	30 000	1500	900

change of configuration from *cis* to *trans*; (vi) it gives products which are similar in type but not identical in structure to those obtained by autoxidation; and (vii) it is a quicker reaction than autoxidation, especially for monoene esters (Table 7.3), and its rate is related to the number of double bonds rather than the number of doubly activated allylic groups.

7.2.4 Decomposition of hydroperoxides to short-chain compounds

This topic is discussed before detailing the structure of the hydroperoxides so that for individual acids and esters it is possible to discuss the hydroperoxides and their breakdown products together.

Short-chain (volatile) compounds which include aldehydes, ketones, alcohols, acids, esters, lactones, ethers, and hydrocarbons contribute to odour and flavour. Sometimes this is considered to be favourable, imparting characteristic and acceptable flavours to raw (mainly by enzymic changes) and cooked food (usually non-enzymic). Sometimes these changes produce the undesirable flavours and odours associated with rancid fat.

Aldehydes are particularly significant in flavour studies. They result from hydroperoxide decomposition by a homolytic or heterolytic mechanism of which the former is more common (Figure 7.5). Each hydroperoxide

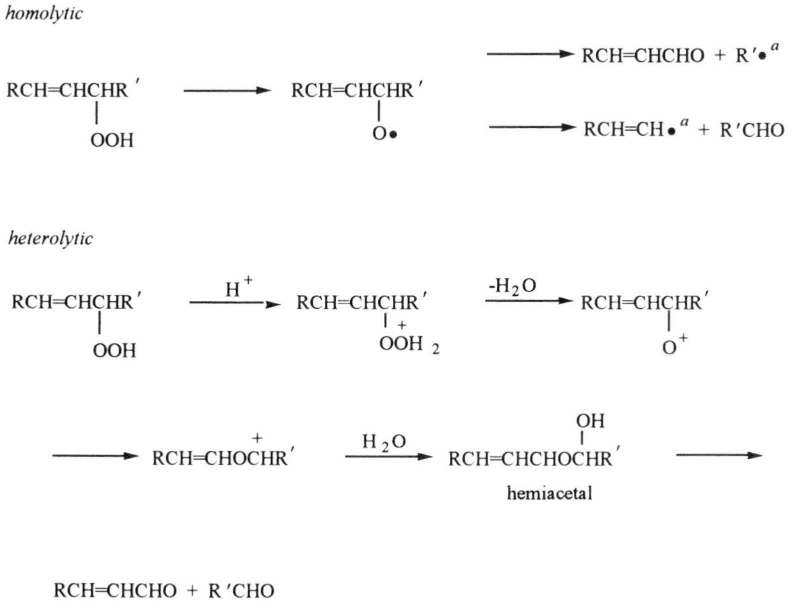

Figure 7.5 Homolytic and heterolytic breakdown of allylic hydroperoxides. [a] These species are converted to hydrocarbons by hydrogen abstraction or to alcohols by reaction with •OH.

can produce two aldehydes of which the short-chain volatile member is more significant. The other aldehyde is also attached to an ester function. In a glycerol ester these remain in the oil/fat and are described as core aldehydes. Aldehydes produced from natural fats are complex mixtures because of the large number of hydroperoxides from which they can be produced. Details are discussed later but these are summarized in Tables 7.5 and 7.6. With partially hydrogenated fats the number of aldehydes which can be produced is still greater because of the large number of double bond positions in such compounds.

Most of these aldehydes have a very low threshold value so they need be present only at minute levels to exert their olfactory effect. For example, the 9-hydroperoxide from linoleate gives 2,4-decadienal which has a deep-fried flavour at a concentration of 5×10^{-10} M (i.e. 0.5 parts per billion). Some other very low threshold values are cited in Table 7.4.

It is still not possible to provide a satisfactory explanation of all the short-chain compounds which have been identified. These could arise from unrecognized hydroperoxides by conventional reactions, from known hydroperoxides by unidentified pathways of breakdown, or from unidentified minor acids in the original oil. Many of them probably result from the further oxidation of unsaturated aldehydes and other decomposition products.

The formation and decomposition of hydroperoxides in aqueous solution under physiological conditions furnishes a number of short-chain polyfunctional compounds such as the very reactive C_8 and C_9 4-hydroperoxy-2-enals:

$$CH_3(CH_2)_nCH(OOH)CH=CHCHO$$
$$n = 3 \ (C_8) \text{ or } 4 \ (C_9)$$

Table 7.4 Threshold flavour values of some aldehydes and their sources from oxidized fats

Aldehyde	Source of hydroperoxide	Threshold value (ppb)
5:1 (2)	18:3	46
6:1 (3)	18:3	90
7:1 (4c)	cod muscle	0.5–1.6
7:2 (2t4c)	18:3	55
9:1 (6t)	18:2 (9,15)	0.3
9:2 (3c6c)	18:3	1.5
9:2 (2t6c)	18:3	2
10:3 (2t4c7c)	18:3	150

Also 1,5c-octadien-3-one (butter fat) 0.02 ppb.

Table 7.5 Major hydroperoxides produced from methyl oleate, linoleate, and linolenate by autoxidation and photo-oxygenation[a]

Methyl ester	Autoxidation			Photo-oxidation		
	OOH[b]	db[b]	yield (%)	OOH[b]	db[b]	yield (%)
oleate	8	9 (c/t)	27	–		
	9	10 (mainly t)	24	9	10t	50
	10	8 (mainly t)	23	10	8t	50
	11	9 (c/t)	26	–		
linoleate	9	10t12c[cd]	52	9	10t12c[d]	32
	–			10	8t12c[e]	17
	–			12	9c13t[e]	17
	13	9c11t[cd]	48	13	9c11t[d]	34
linolenate	9	10,12,15[d]	32	9	10,12,15[d]	23
	–			10	8,12,15[e]	13
	12	9,13,15[de]	11	12	9,13,15[de]	12
	13	9,11,15[de]	11	13	9,11,15[de]	14
	–			15	9,12,16[e]	13
	16	9,12,14[d]	46	16	9,12,14[d]	25

[a]Yields are based on total mono-hydroperoxides only: other more highly oxygenated compounds may also be present.
[b]Carbon atoms for hydroperoxide and double bond(s).
[c]Other hydroperoxides (8, 10, 12, 14 but not 11) are present as minor components (0.5–1.5% each).
[d]Contain conjugated diene systems.
[e]Can form a cyclic peroxide (see text).

7.2.5 Reaction products

The more important hydroperoxides obtained from methyl oleate, linoleate, and linolenate and the short-chain compounds produced from them are summarized in Tables 7.5 and 7.6.

(a) *Methyl oleate.* The oxidation of methyl oleate has been extensively studied since this is considered to be the prototype for all monoene esters. Photo-oxidation produces only the 9- and 10-hydroperoxides in equal amounts with the double bond having changed position and configuration. Pure methyl oleate undergoes autoxidation at a modest rate only after a long induction period. The situation is less simple in oleic-rich oils since these may contain pro-oxidants and antioxidants as well as linoleic esters which will oxidize more quickly and then promote the oxidation of oleic esters. Since reaction can involve either allylic centre and proceed through resonance-stabilized radicals the product may contain eight different compounds:

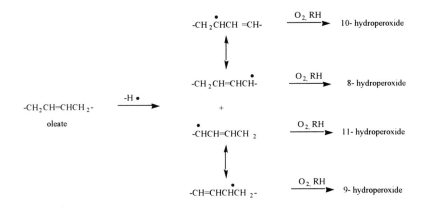

Table 7.6 Short-chain compounds expected from breakdown of hydroperoxides derived from methyl oleate, linoleate, and linolenate[a]

Type of compound	Oleate	Linoleate	Linolenate
aldehydes	8:0	6:0	3:0
	9:0	7:1 (2*t*)	4:1 (2*t*)
	10:0	9:1 (3*c*)	6:1 (3*c*)
	10:1 (2*t*)	10:2 (2*t*4*c*)	7:2 (2*t*4*c*)
	11:1 (2*t*)		9:2 (3*c*6*c*)
			10:3 (2*t*4*c*7*c*)
hydrocarbons	7:0	5:0	2:0
	8:0	8:1 (2)	5:1 (2)
			8:2 (2,5)
alcohols	7:0	5:0	2:0
	8:0	8:1 (2)	5:1 (2)
			8:2 (2,5)
methyl esters	7:0	8:0	8:0
	8:0	11:1 (9)	11:1 (9)
			14:2 (9,12)
ω-oxo esters	8:0	9:0	9:0
	9:0	10:1 (8)	10:1 (8)
	10:0	12:1 (9)	12:1 (9)
	10:1 (8)	13:2 (9,11)	13:2 (9,11)
	11:1 (9)		15:2 (9,12)
			16:3 (9,12,14)

[a]Most, but not all, of these compounds have been observed.
Source: Frankel (1980), *Prog. Lipid Res.*, **19**, 1.

Among the short-chain compounds resulting from these hydroperoxides the aldehydes are probably the most important. The 8-, 9-, 10-, and 11-hydroperoxides should give 11:1, 10:1, 9:0, and 8:0 aldehydes along with a similar series (8:0–11:1) of ω-oxo esters.

Oleic acid reacts with selenium dioxide/tertbutylhydroperoxide to give a mixture of 8-hydroxy, 11-hydroxy- and 8,11-dihydroxy derivatives of elaidic acid (9*t*-18:1). The corresponding mono and dihydroxy stearates are produced from them by reduction with hydrazine/air.

(b) *Methyl linoleate.* This ester gives the expected four hydroperoxides on photo-oxidation (Table 7.5). The 9- and 13-hydroperoxides are designated as outer hydroperoxides and the 10- and 12-hydroperoxides (which are produced only by photo-oxidation) are inner hydroperoxides. Only the two outer hydroperoxides contain a conjugated diene unit. The two inner hydroperoxides are present at lower levels among the total hydroperoxides and this is thought to result from the presence of an alternative reaction path which leads to a hydroperoxy peroxide. This is not possible with the outer hydroperoxides which do not contain the required β-hydroperoxy ene unit.

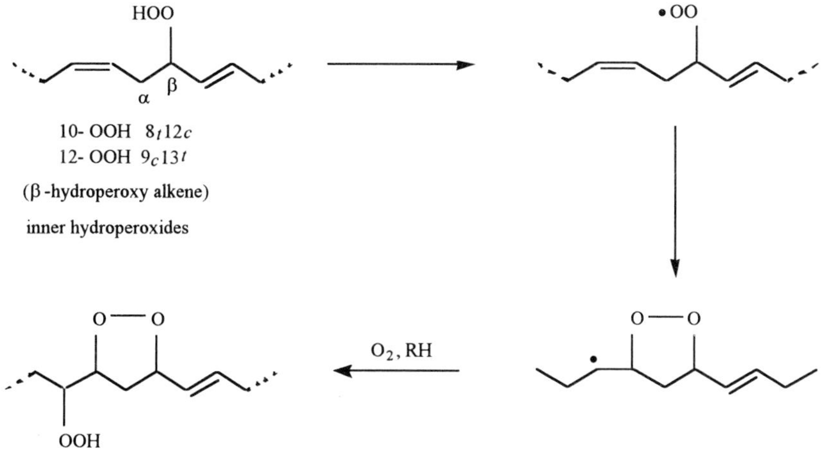

HOO

α β

10- OOH 8$_t$12c
12- OOH 9$_c$13t

(β -hydroperoxy alkene)

inner hydroperoxides

•OO

O — O

OOH

O — O

O$_2$, RH

hydroperoxy peroxides

In contrast, autoxidation gives only two major products in roughly equal amounts, although these may be accompanied by minor products resulting from reaction initiated at the C8 and C14 allylic centres (see Table 7.5). There is no evidence of any 11-hydroperoxide. The *ct* dienes first formed may change to *tt* dienes with exchange of the hydroperoxide function from C9 to C13 or vice versa (see top of p. 170).

Autoxidation of linoleate is 20–40 times quicker than autoxidation of oleate. This relates to the greater reactivity of the doubly allylic CH$_2$ group (C11) which does not exist in oleate. Reaction is further accelerated as the methylene-interrupted polyene system is extended and, to a first approximation, the rate of autoxidation relative to linoleate (=1) is equivalent to

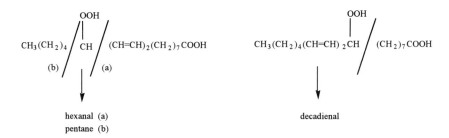

the number of doubly allylic CH_2 groups and, for example, EPA (20:5) and DHA (22:6) would be expected to oxidize approximately four and five times as quickly as linoleate. Non-methylene-interrupted dienes, on the other hand, react more like oleate than linoleate and this is probably related to the absence of a doubly allylic CH_2 group in such acids.

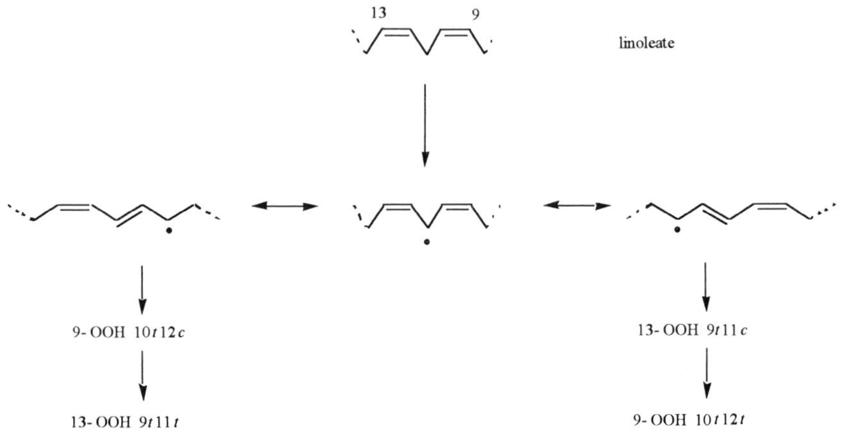

Short-chain compounds resulting from linoleate hydroperoxides are listed in Table 7.6. The most important aldehydes are hexanal from the C13 hydroperoxide and decadienal from the C9 hydroperoxide. The measurement of pentane by head space analysis has also been used as an index of linoleate oxidation. It can be detected in exhaled breath and used as a measure of *in vivo* oxidation.

$CH_3(CH_2)_4$ CH $\overset{\text{OOH}}{|}$ $(CH=CH)_2(CH_2)_7COOH$

(b) (a)

hexanal (a)
pentane (b)

$CH_3(CH_2)_4(CH=CH)_2CH$ $\overset{\text{OOH}}{|}$ $(CH_2)_7COOH$

decadienal

(c) *Methyl linolenate.*   Methyl linolenate contains two pentadiene units (9,12 and 12,15). The monohydroperoxides produced from linolenate by photo-oxidation (six) and autoxidation (four) are listed in Table 7.5. In both cases outer hydroperoxides, which cannot give cyclic peroxides, exceed the inner hydroperoxides, which can. Four of the hydroperoxides still contain a pentadiene unit (9-HPO-10,12,15, 10-HPO-8,12,15, 15-HPO,9,12,16, and 16-HPO-9,12,14) and these can be oxidized further by photo-oxidation or autoxidation to give bishydroperoxides.

Among important short-chain compounds from linolenate (and from other *n*-3 polyene acids) are propanal, hexenal, nonadienal, and decatrienal which come from the 16-, 13-, 10-, and 9-hydroperoxides respectively. Two C_8 ketones (see Table 7.4) and their corresponding alcohols come from the 9-hydroperoxy-10,12-peroxy acid. The aldehydes from linolenate generally have lower threshold values than those from linoleate and this, along with the quicker oxidation of the triene ester, accounts for the rapid development of rancid flavour in oils containing linolenate and other *n*-3 acids (Table 7.4).

(d) *Cholesterol.* The possible link between cholesterol oxidation products and coronary heart disease and other disease states make it appropriate to add a brief note on the oxidation of cholesterol. In the solid phase oxidation has been observed at the tertiary carbon atoms C20 and C25 in the side chain but this is not important in solution or in aqueous dispersion when, as expected, oxidation is related to the double bond (Δ5) and its associated allylic centres. Photo-oxidation gives the 5-hydroperoxide but autoxidation gives the 7-hydroperoxide as the major primary product. This furnishes the 7-hydroxy derivative (a 3,7-diol) or the 7-oxo compound. This last is unstable in alkaline solution and is converted to the 3,5-dien-7-one. The oxidation products also include a 5,6-epoxide which is the precursor of the 3,5,6-triol. The epoxide is not a product of direct oxidation but results from interaction of cholesterol with hydroperoxide. A typical autoxidation product was shown to contain the following components: 7-keto (47%), 7-hydroxy (18%), epoxy and trihydroxy (5%), and unreacted cholesterol (30%). Full stereochemical details of the oxidation products are given by Maerker (1987) (See top of p. 172).

7.2.6 *Other secondary reaction products*

Hydroperoxides produced by autoxidation or photo-oxidation are readily converted to a variety of compounds with the same chain length. These may result from homolytic or heterolytic processes and in some cases

cholesterol

involve reaction with water or other solvent. A selection of typical stuctures based on linoleate hydroperoxide are given. These are to be read as partial structures representing C8–C14. The hydroxy diene, for example represents both the 9- and 13-hydroxy compounds.

α- and γ- ketols (Z = OH, OMe, SEt, OCOR)

(also the corresponding triols)

Hydroperoxides and their more extensively oxidized derivatives also react to form dimers and compounds of still higher molecular weight. Such compounds can be separated and measured by gel permeation HPLC. Processed fish oils generally contain about 2% of such materials but this value rises to around 30–40% during oxidation at 35°C in air and light over 4 days. The products, whether from fish oils or from C_{18} polyene acids are likely to be complex mixtures and the structure of such compounds has not been fully identified but it is known that there are dimers linked through peroxide (C–O–O–C), ether (C–O–C), or hydrocarbon (C–C) bridges with the first of these predominating in oxidation at around 40°C. A typical partial structure is shown below. It will be accompanied by hydroperoxy and peroxy derivatives.

16 (8)

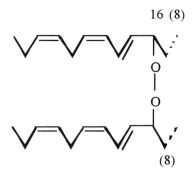

(8)

peroxide linked dimer of linolenic acid

7.3   Antioxidants

7.3.1   Introduction

All olefinic compounds are liable to oxidative deterioration and the successful everyday use of such materials is generally dependent on the inhibition of this change through the use of antioxidants. These are added to rubber goods, plastics, and lubricants, for example, as well as to oils and fats and foods. They are also important *in vivo* since oxidative deterioration leads to a number of undesirable disease conditions.

Antioxidants which can be added to fats and to fatty foods are rigorously controlled. Only permitted substances may be used and then only below certain agreed levels. The matter is complicated further by the fact that not all these substances are accepted in all countries. For example TBHQ (section 7.3.2) is allowed in the USA but not in countries of the EU. Antioxidants permitted in Europe have E numbers assigned to them.

Antioxidants can be grouped according to the ways in which they operate (sections 7.3.2–7.3.4) and can also be classified as natural or synthetic. The synthetic compounds are generally cheaper and frequently more effective than the natural materials. Nevertheless, there is a move away from synthetic antioxidants and an increasing demand for natural materials. Much of the large and increasing processed food industry would be impossible without antioxidants of some kind. They are essential to inhibit the quick development of rancidity.

Vegetable oils generally contain natural antioxidants, i.e. nature provides its own safeguards. These are extracted along with the oil though their level may be reduced during refining. These compounds are usually tocols (Figure 7.7), i.e. tocopherols and tocotrienols, although other natural phenolic compounds also occur widely (Figure 7.8). Data for some of the tocols are given in Table 7.7. It is clear that vegetable oils are generally good sources of tocols. Some are particularly rich and soybean oil and palm oil are important sources of natural tocol mixtures generally recovered from minor fractions produced during refining. Fish oils and animal fats are much less well protected. Therefore they have a greater need for added antioxidant and the effect of this is much more marked than with similar additions to vegetable oils which already have some in-built protection.

Table 7.7 Content of tocols in vegetable oils (mg/kg ≡ ppm)[a]

Oil	Tocopherols				Tocotrienols	
	α	β	γ	δ	α	β
canola	210	1	42	–	–	–
coconut	5	–	–	6	5	–
corn	112	50	602	19	–	–
cottonseed	389	–	387	–	–	–
olive	119	–	7	–	–	–
palm	256	–	316	70	146	3
palm kernel	62	–	–	–	–	–
peanut	130	–	214	21	–	–
safflower	342	–	71	–	–	–
sesame	136	–	290	–	–	–
soybean	75	15	797	266	2	–
sunflower	487	–	51	8	–	–
walnut	563	–	595	450	–	–
wheatgerm	1330	710	260	271	26	18
cod liver	220	–	–	–	–	–
herring	92	–	–	–	–	–
menhaden	75	–	–	–	–	–
lard	12	–	7	–	7	–
tallow	27	–	–	–	–	–

Source: Adapted from *The Lipid Handbook*, 2nd edn (1994) p. 131.
[a]These are typical values: the observed ranges are quite wide. Values below 1 ppm are shown as –.

The mixed tocols are also known as vitamin E but vitamin activity is not the same as antioxidant activity and the various tocols are not equally effective as antioxidants and as vitamins (see section 3.9).

Attempts to assess and compare antioxidants are complicated by several factors and the following points should be noted. (1) Effects vary with different oils and fats because of the different types and levels of unsaturated acids present and because of differing levels of antioxidant already present. (2) Different results are obtained at different temperatures because mechanisms of oxidation and of hydroperoxide-breakdown change with temperature and also antioxidants differ in their volatility at elevated temperatures. (3) Results vary with the test procedure which may measure primary oxidation products (hydroperoxides) or secondary oxidation products (carbonyl compounds and/or volatile compounds). (4) Mixtures of antioxidants are frequently employed and these often act synergistically, i.e. a mixture is more effective than would be expected from activity measured on the individual members of the mixture. (5) Solubility factors may also be important, especially in two-phase systems where antioxidants may be distributed between aqueous and lipid phases.

The major antioxidants will now be described and will be grouped according to their mode of action which relates to the mechanisms of oxidation (sections 7.2.2 and 7.2.3).

7.3.2 Primary antioxidants (chain-breaking)

Antioxidants in this category reduce the propagation cycle by promoting termination processes. These are mainly phenols or amines or compounds with extensive conjugated unsaturation such as carotenoids. They react with peroxy radicals to give products which will not sustain the propagation sequence:

$$ROO\bullet + AH \longrightarrow ROOH + A\bullet \; -----\rightarrow \quad products$$

$$ROO\bullet + B \longrightarrow RO\overset{\bullet}{OB} \qquad -----\rightarrow \quad products$$

AH = amines or phenols   B = β-carotene etc.

Antioxidants act in a sacrificial manner and the induction period ends when all or most of the antioxidant has been used up. As will be indicated later it is sometimes possible to regenerate antioxidants and so extend their usefulness. Also some antioxidants react with two peroxy radicals and sometimes compounds produced from the antioxidant themselves show further antioxidant activity until in turn they are also fully used up. The antioxidants produce a wide range of reaction products but these are mainly dimers or substitution products such as these formulated in Figures 7.6 and 7.7.

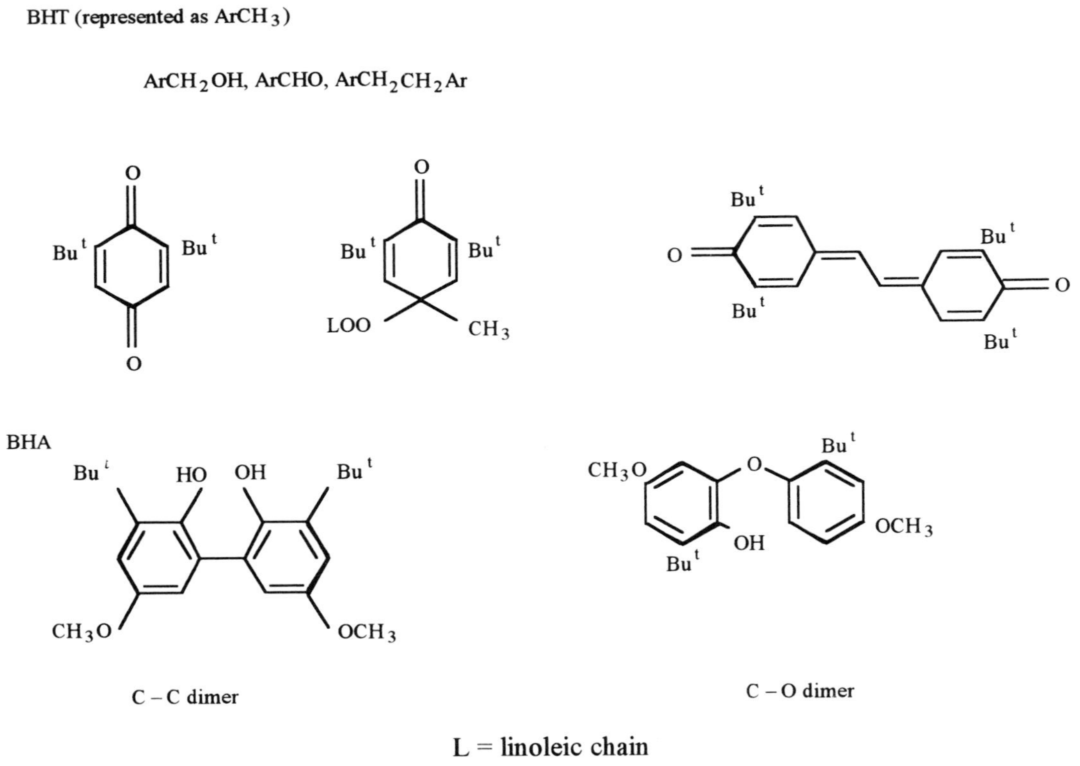

Figure 7.6 Structures of compounds produced from antioxidants by their oxidation.

There are four important synthetic phenolic antioxidants: butylated hydroxyanisole (BHA), butylated hydroxytoluene (BHT), gallic esters especially propyl (PG), and tertbutyl hydroquinone (TBHQ) with the E numbers shown:

BHA
(E230)

BHT
(E321)

PG
(E310)

TBHQ
(no E number)

not a permitted food
ingredient in Europe

The phenols (ArOH) are converted to radicals (ArO•) which are stabilized by extensive delocalization of the odd electron.

BHA shows good solubility in fat and reasonable stability in fried and baked products. It is very effective with animal fats and somewhat less so with vegetable oils. It shows marked synergism with both BHT and PG and can be used to a maximum level of 200 ppm.

BHT is less soluble than BHA and is not soluble in propylene glycol (sometimes used as a solvent for antioxidants). It is synergistic with BHA but not with PG and can be used to a maximum level of 200 ppm.

PG is less soluble than either BHA or BHT and does not generally survive cooking. Nevertheless it is effective when used with BHA. It may be used up to 100 ppm.

TBHQ can be used in USA but not in EU countries. It is very effective with vegetable oils, has good solubility, and is stable at high temperatures. It is useful for oil storage and can be completely removed during deodorization.

A natural tocol source may contain up to four tocopherols and four related tocotrienols (Figure 7.7 and Table 7.7). Concentrates can be obtained from refining residues and concentrated by molecular distillation. Their composition will vary with the source, sometimes being α-rich, sometimes γ-rich, and sometimes rich in tocotrienols. α-tocopherol has the highest vitamin E activity but shows lower antioxidant activity than the other tocopherols, particularly the δ and γ forms. Natural tocopherol mixtures are usually used at levels up to 500 ppm along with ascorbyl palmitate (200–500 ppm) which further increases the tocopherol activity. At higher levels (>1000 ppm) α-tocopherol is considered to act as a pro-oxidant. Since vegetable oils already contain tocols at levels of 200–800 ppm further additions often show only a limited effect.

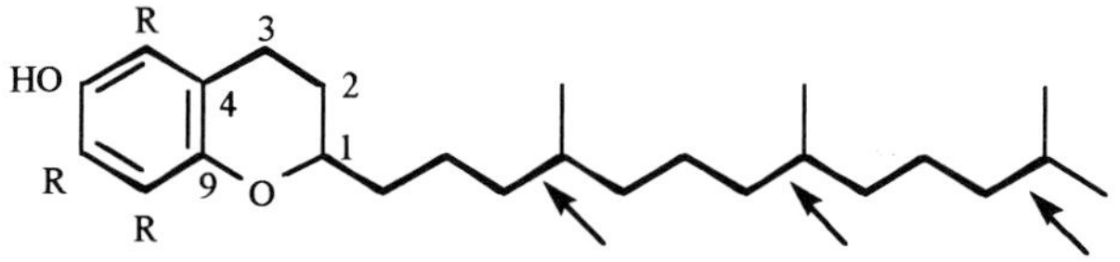

R= H or CH_3

The tocols have 1-3 methyl groups as follows:

α (5,7,8),   β (5,8),   γ (7,8), and   δ (8)

a   The tocotrienols have three double bonds in the C_{16} side-chain in the positions indicated by arrows.

b   Oxidation products of tocopherols

		dimer
α ⟶	Toc CH_2 CH_2 Toc	C-C 5,5'
γ ⟶	Toc-Toc	C-C 5,5'
	Toc-O-Toc	C-O 5,6'
δ ⟶	Toc-Toc	C-C 5,5'
	Toc-O-Toc	C-O 5,6'

also   Toc-O-O-C_{18} chain (C9 or C13) C_{18} chain derived

Toc-O-C_{18} chain (epoxide)   from linoleic acid

c   E number for tocopherols concentrate is E306

Figure 7.7 Structures of tocopherols and their oxidation products, and of tocotrienols[a].

Other natural phenolic antioxidants are present in a wide range of plant sources and the use of herbs and spices to preserve foods predates the concept of shelf-life. Many of them also have a strong flavour which may or may not be desirable depending on circumstances. Rosemary leaves, after steam distillation to remove essential oil, give a crude ethanolic extract with considerable antioxidant activity which is now available for commercial use. It has been shown to contain several phenolic compounds including carnosic acid and rosmaric acid (Figure 7.8).

Other plant extracts with useful antioxidant activity include thyme, sage, myrtle, tea, and oats. Sesame oil contains its own antioxidant (sesamol) and this oil is sometimes added to other oils to act as an antioxidant.

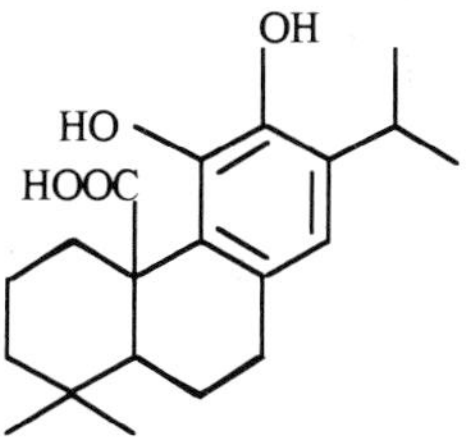

carnosic acid
(rosemary)

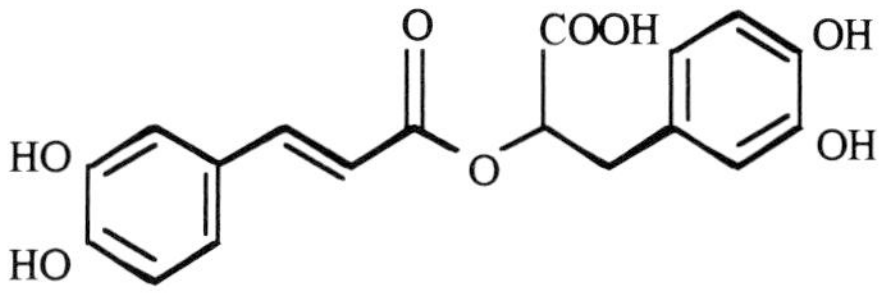

rosmaric acid
(rosemary)

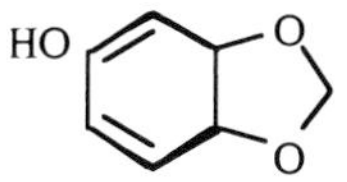

sesamol
(sesame oil)

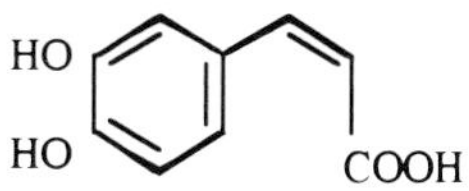

caffeic acid
(soybeans)

Figure 7.8 Some other natural antioxidants.

β-carotene and similar substances containing extensive conjugated unsaturation inhibit photo-oxygenation by their ability to quench singlet oxygen (section 7.2.3) but these compounds can also inhibit autoxidation through their ability to react with and remove peroxy radicals. When this happens the odd electron is delocalized over the conjugated polyene system. Under other conditions β-carotene can act as a pro-oxidant.

7.3.3   Secondary antioxidants

A second group of compounds showing antioxidant activity operate through inhibition of the chain-initiating process. These act mainly through chelation of the metal ions which promote initiation. The concentration of metals required to reduce the keeping time of lard at 98°C by 50% is 0.05 ppm for copper and 0.6 ppm for iron. It is therefore essential to keep these metals at as low a level as possible. Useful metal chelators include ethylene diamine tetra-acetic acid (EDTA), citric acid, phosphoric acid, and certain amino acids. They are often added along with the chain-breaking antioxidants which have already been described.

7.3.4   Other materials which inhibit oxidative deterioration

Still other compounds can be used to enhance antioxidant activity. Vitamin C, for example, is useful because it interacts with spent tocopherols (vitamin E) to regenerate these. Thus vitamin C shows a sparing activity. However, vitamin C is a water-soluble compound with very low lipid solubility and is therefore often used in the form of ascorbyl palmitate which is more soluble in fat. Phospholipids also promote antioxidant activity though it is not clear how far this results from chelation of metal ions and how far they act as emulsifying agents bringing antioxidant and fat together.

All these materials are concerned with inhibition of the chain-radical autoxidation reaction and have little or no effect on photo-oxygenation occurring by a different reaction mechanism (section 7.2.3). This process is inhibited by singlet oxygen quenchers of which the best known is β-carotene and other carotenoids. These may be present in natural vegetable oils and if not can be added at levels of 5–10 ppm. Not all carotenes are equally effective. Activity increases with the length of the conjugated polyene system and seems to be enhanced by the presence of keto groups.

It is important to recognize that antioxidants do not prevent oxidation. They extend the inhibition period during which time oxidation is very slow and of no major consequence. It follows that appropriate antioxidants should be added as soon as the oil/fat is extracted and before oxidation has started. No amount of antioxidant can regenerate a fat which is already oxidized. Also conditions which promote oxidation should be avoided as far as possible. This involves avoiding elevated temperatures, unnecessary contact with air (nitrogen blanketing when possible and avoid splashing which will increase air solubility), and unnecessary exposure to light. Storage should always be under the best conditions and exposure to iron and copper in storage vessels and pipe lines must be avoided.

7.4 Biological oxidation

7.4.1 α-Oxidation

Although less common than β-oxidation, α-oxidation is an important pathway, especially when β-oxidation is blocked, and leads to α-hydroxy acids and to acids with one less carbon atom (*nor*-acids). The latter generally have an odd number of carbon atoms. Reaction occurs via the α-hydroperoxy acid and the *nor*-aldehyde thus:

$$RCH_2COOH \xrightarrow{O_2} RCH(OOH)COOH \xrightarrow[-H_2O]{-CO_2} RCHO \longrightarrow RCOOH$$

$$RCH(OOH)COOH \downarrow$$

$$RCH(OH)COOH$$

α-hydroxy acids are important components of sphingolipids and also occur in wool wax. The seed oil of *Salva nilotica* contains the acids listed in

Figure 7.9 and it seems that each of these is linked with a common C_{18} acid through α-oxidation. 2-hydroxysterculic acid also occurs naturally and is probably the link between the two common cyclopropene acids – sterculic (C_{19}) and malvalic (C_{18}) (Figure 7.10).

7.4.2   β-Oxidation

Superficially, β-oxidation appears as a reversal of the biosynthetic route by which long-chain acids are produced. It involves a number of steps and each cycle of reactions gives a product with two fewer carbon atoms. This leads eventually to the complete degradation of saturated acids except that odd chains finally produce a molecule of propionate rather than acetate. Unsaturated acids behave similarly except that an isomerase may be required to get double bonds into the appropriate position and configuration (Figure 7.11).

7.4.3   ω-Oxidation

Long-chain acids are sometimes oxidized at the ω-end of the chain. The product is a dibasic acid produced via the ω-hydroxy acid (also naturally

C_{18} acid	oleic	linoleic	linolenic
	↓	↓	↓
hydroxy acid	2-hydroxyoleic	2-hydroxylinoleic	2-hydroxylinolenic
	↓	↓	↓
C_{17} acid	17:1 (8c)	17:2 (8c,11c)	17:3 (8c,11c,14c)

Figure 7.9 Some acids present in *Salva nilotica* seed oil linked through α-oxidation.

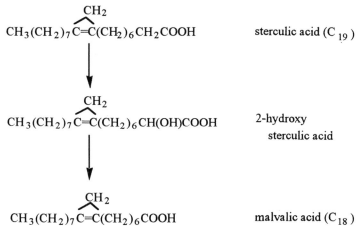

Figure 7.10 Some cyclopropene acids linked through α-oxidation.

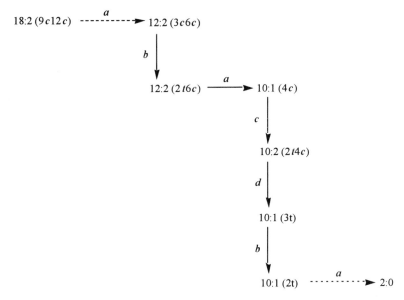

Figure 7.11 β-oxidation of linoleic acid (as coenzyme A ester). *a*, β-oxidation; *b*, enoyl-CoA isomerase; *c*, acyl-CoA dehydrogenase; *d*, 2,4-dienoyl-CoA reductase.

occurring). The dibasic acid may then be subjected to β-oxidation at the ω end of the molecule and this happens particularly when normal β-oxidation at the carboxyl end is hindered.

7.4.4 Lipoxygenase

The enzymes which catalyse reaction between *cis, cis*-1,4-dienes and oxygen are known as lipoxygenases. They occur widely in the plant and animal kingdom, and particularly in the former. The best substrate seems to be linoleic acid which is oxidized to the 9- and/or 13-hydroperoxide. These hyperoxides suffer fragmentation through a hydroperoxide lyase to give short-chain compounds some of which have marked and characteristic odours (Figure 7.12). These, and similar products from linolenic acid, are also converted to a wide range of C_{18} compounds including hydroxy dienes, vinyl ethers, α- and γ-ketols, and 12-oxophytodienic acid.

12- oxophytodienic acid

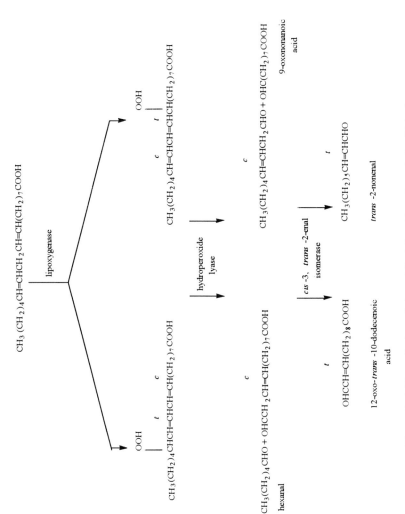

Figure 7.12 Enzymic degradation of linoleic acid to short-chain aldehydes.

The mechanism of enzymic oxidation is given in Figure 7.13.

7.4.5 Production and function of eicosanoids

Arachidonic acid (20:4 *n*-6), liberated from appropriate phospholipids by phospholipase A2, is the precursor of a wide range of enzymic oxidation products. Related compounds from 20:3 (*n*-6) and 20:5 (*n*-3) are also known but have not been so extensively studied. Many of these products are formed in animal systems on demand, have a very short half-life (seconds or minutes), and show marked physiological activity so that they behave almost like hormones. These include a range of C_{20} compounds with a cyclopentane ring linking C8 to C12 (e.g. prostaglandins) and also many acyclic C_{20} acids containing hydroperoxy, hydroxy, dihydroxy, or epoxy functions (Tables 7.8 and 7.9). In forming these compounds double bonds may disappear or may change position and configuration. Some structures are given below.

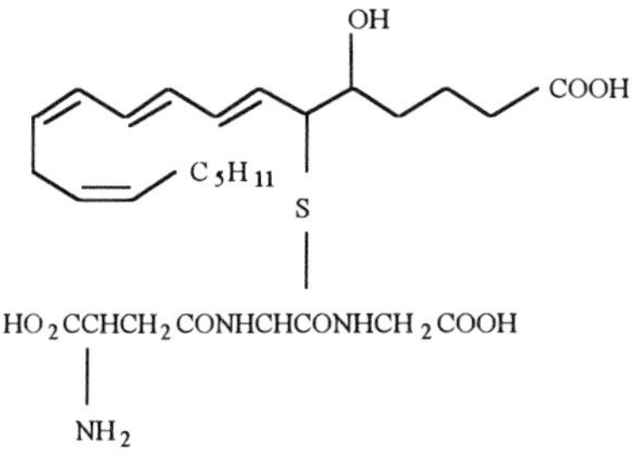

leukotriene C_4

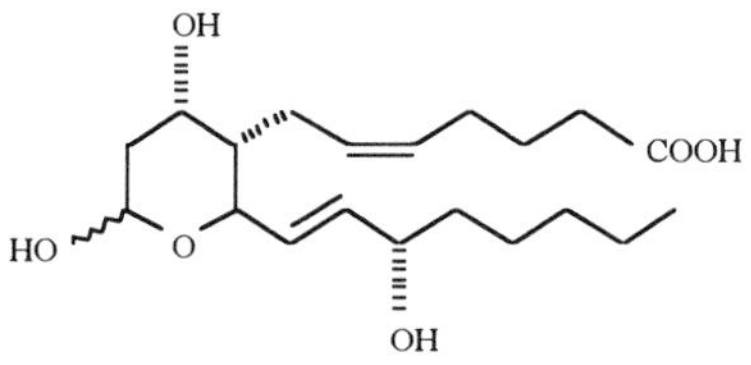

TXB $_2$ (thromboxane)

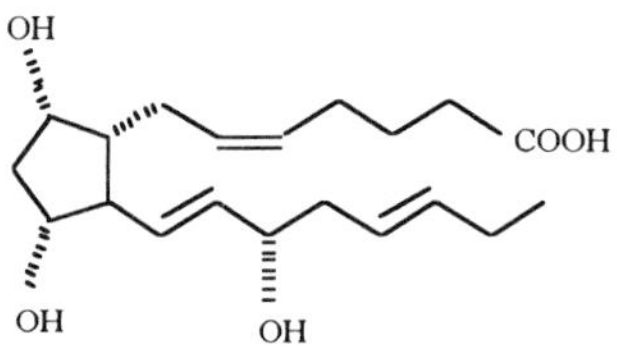

PGF $_2 \alpha$ (prostaglandin)

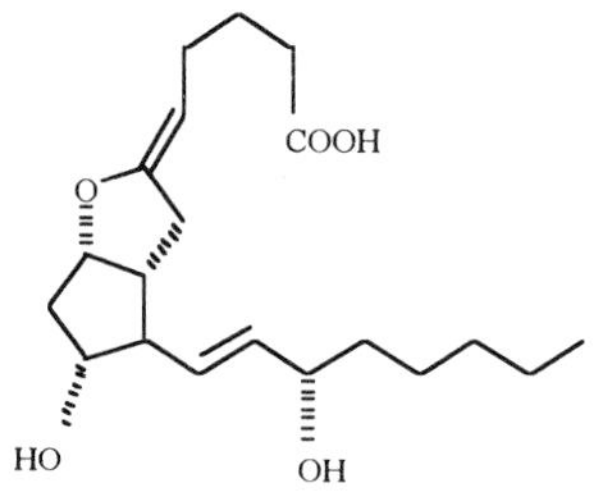

PGI $_2$ (prostacyclin)

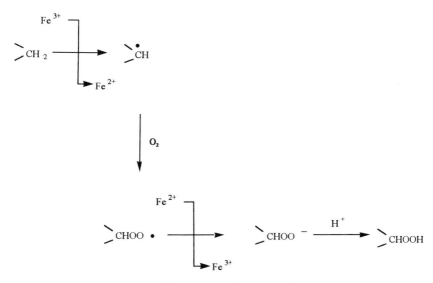

Figure 7.13 Mechanism of enzymic oxidation.

Table 7.8 Some reactions in the eicosanoid cascade

Enzyme	Reaction	Primary products	Secondary products
cyto P450	ω-hydroxylation	hydroxy acids	oxo acids
cyto P450	epoxidation	EpETE	
lipoxygenase	hydroperoxidation	HpETE	HETE, DHETE
			lipoxins
			hepoxilins
			LT
cyclo-oxygenase	oxidative	PGG	PG, TX,
	cyclization	PGH	HHT, PGI

Abbreviations: see Table 7.9 and structures in text.
HHT = 12(*S*)-hydroxy 17:3 (5*c*8*t*10*t*).

Table 7.9 Typical products of the eicosanoid cascade

Abbreviation	Name (acid)	Example (structure)
HpETE	hydroperoxyeicosatetraenoic	5-OOH 20:4 (6*t*8*c*11*c*14*c*)
HETE	hydroxyeicosatetraenoic	5-OH 20:4 (6*t*8*c*11*c*14*c*)
DHETE	dihydroxyeicosatetraenoic	5,6-diOH 20:4 (7*t*9*t*11*c*14*c*)
		5,15-diOH 20:4 (6*t*8*c*11*c*13*t*)
lipoxins	trihydroxyeicosatetraenoic	5,6,15-triOH 20:4 (7*t*9*t*11*c*13*t*)
		5,14,15-triOH 20:4 (6*t*8*c*10*t*12*t*)
EpETE	epoxyeicosatrienoic	5,6-ep 20:3 (8*c*11*c*14*c*)
EpOME	epoxyoctadecenoic	9,10-ep 18:1 (12*c*)[a]
		12,13-ep 18:1 (9*c*)[a]

[a]C_{18} epoxy acids occur naturally as coronaric and vernolic acid, respectively (section 1.12.2).

7.5 Other oxidation reactions

7.5.1 Epoxidation

Epoxidation is the name given to the reaction where a double bond is converted into an epoxide, i.e. a three-membered cyclic ether.

$$-CH=CH- \longrightarrow \overset{\overset{\displaystyle O}{\displaystyle \wedge}}{-CHCH-}$$

Long-chain epoxy acids which occur naturally are discussed in section 1.12.2.

(a) *Preparation – methods, mechanism, and products.* Epoxidation is an important process carried out on a laboratory scale and as an industrial process. Epoxidized oils, produced at a level around 50 000 tonnes per year, are used mainly as plasticizers and stabilizers for PVC. Some other uses are indicated in section 7.5.1c. The epoxidation of vegetable oils is reported to provide the largest single use of hydrogen peroxide.

Epoxidation usually occurs through reaction with a peroxy acid which is often prepared *in situ*. These acids result from the reaction of carboxylic acid or other acyl compound with hydrogen peroxide in the presence of an acidic catalyst when necessary:

$$RCOOH + H_2O_2 \overset{H^+}{\rightleftharpoons} RCO_3H + H_2O$$

$$(RCO)_2O + H_2O_2 \longrightarrow RCO_3H + RCO_2H$$

$$RCOCl + H_2O_2 \longrightarrow RCO_3H + HCl$$

Peroxyformic or peroxyacetic acid are widely used on an industrial scale but other peroxy acids are sometimes more convenient in the laboratory. They include peroxytrifluoroacetic, peroxylauric, peroxybenzoic, 3-chloroperoxybenzoic acid, and the monoperoxy acids derived from succinic, maleic, or phthalic anhydride. 3-chloroperoxybenzoic acid and monoperoxyphthalic acid can be safely stored at 0–20°C and are convenient for small-scale reactions. The epoxidation reaction is exothermic and high concentrations of peroxy acids should be avoided. The epoxides themselves are also reactive substances especially in acidic solutions.

The mechanism of the reaction is not finally settled but Bartlett's proposal continues to find favour (Figure 7.14).

Epoxidation is a *cis* addition process so that oleic and elaidic acid give *cis*- and *trans*-epoxystearic acid, respectively. Linoleic gives two racemic *cis*, *cis*-diepoxystearic acids which have been separated by crystallization (Table 7.10). Polyunsaturated acids can also be converted to unsaturated

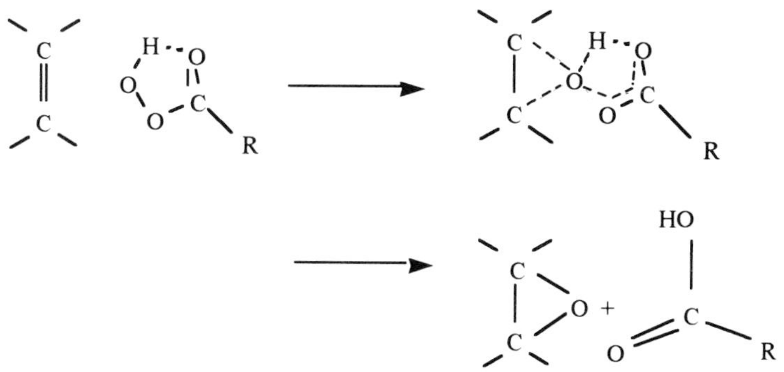

Figure 7.14 Bartlett's mechanism for epoxidation.

Table 7.10 Synthetic epoxy acids

Epoxy acid	Source	mp (°C)
cis-9,10-epoxystearic	oleic	59.5–59.8
trans-9,10-epoxystearic	elaidic	56–57
cis-9,10-cis-12,13-diepoxystearic	linoleic	–
cis-9,10 18:1 (12c)	linoleic	–
cis-12,13 18:1 (9c)	linoleic	–

epoxides by partial epoxidation. For example, three diene monoepoxides can be obtained from linolenic acid.

Attempts to prepare epoxides in other ways include reaction with tert butyl or cumyl hydroperoxide in the presence of molybdenum oxide on alumina or with oxygen and an appropriate aldehyde.

Allenic epoxides are important intermediates in the biological transformation of allylic hydroperoxides:

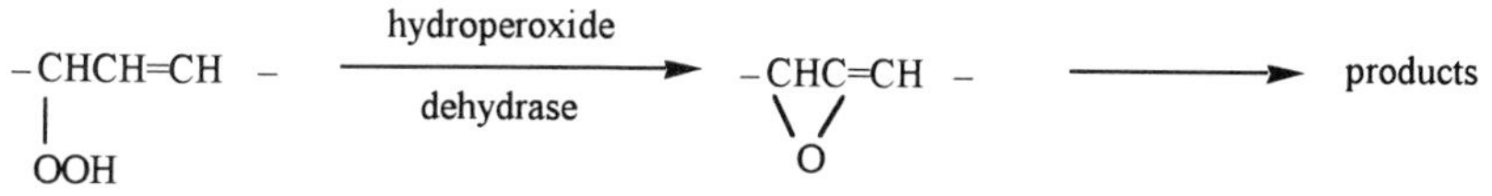

(b) *Spectroscopic and chemical properties.* Epoxy compounds can be recognized and identified by spectroscopic procedures. Reference has already been made (section 6.7.1) to their use as such or after further derivatization in mass spectroscopy. They also show characteristic signals in their NMR spectra and significant chemical shifts are given in Table 7.11.

The epoxide function is very reactive – especially in acidic solution which promotes ring-opening by protonation of the oxygen atom – and interacts with a wide range of nucleophilic reagents:

$$\underset{\text{-CHCH-}}{\overset{\text{O}}{\wedge}} \quad \xrightarrow{\text{H}^+} \quad \underset{\text{-CHCH-}}{\overset{\overset{+}{\text{OH}}}{\wedge}} \quad \xrightarrow{\text{HX}} \quad \text{-CH(OH)CHX-}$$

Reagent (product): HX = H_2 (alcohol), H_2O (diol), ROH (alkoxy alcohol), RCOOH (acyloxy alcohol), $RCONH_2$ (acylamino alcohol), H_2S (mercapto alcohol), R_2NH (dialkylamino alcohol), HCN (cyano alcohol), HBr (bromo alcohol).

Reaction with acetic acid followed by alkaline hydrolysis of the hydroxy acetate is a simple and much-used way of converting epoxides to diols (section 7.5.2) and reaction with hydrogen bromide is the basis of the classical Durbetaki method of determining epoxide content.

Reactions of epoxides (especially from triacylglycerols) with polyhydric alcohols give polyhydroxy compounds which can be reacted with di-isocyanates to give polyurethanes.

Epoxides are converted to ketones by reaction with sodium iodide in polyethyleneglycol. They can also be changed back to alkenes by appropriate stereospecific procedures (see Figure 7.15).

(c) *Applications.* As already indicated epoxy oils are widely used as stabilizers and plasticizers for PVC. Vernolic acid, a major component of *Vernonia galamensis* and of some other seed oils can be used to produce the C_8, C_9, and C_{12} dibasic acids and attempts are being made to produce

Table 7.11 Spectroscopic properties (NMR) of some epoxy acids

Partial structure	Configuration	δH	δC
	cis	2.70 (a) 1.43 (b)	56.67 (a) 27.48 (b) 26.25 (c) 28.85 (d)
	trans	2.45 (a) 1.43 (b)	58.54 (a) 31.89 (b) 25.80 (c) 28.98 (d)
	cis, cis	1.60 (a) 2.91 (b)	–
		2.79 (c) 1.45 (d)	
	cis, cis	1.45 (a) 2.73 (b)	27.40, 27.79 (a, e)[a] 56.40, 57.05 (b)[a]
		2.1–2.3 (c)	26.33, 26.28 (c, f)[a] 132.40, 124.05 (d)[a]
		5.41 (d) 2.03 (e)	

[a]Assignments uncertain between these pairs.

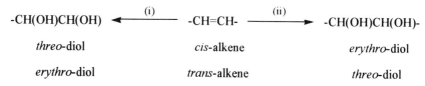

Figure 7.15 The conversion of epoxide to alkenes. Reagents: (i) NaI, NaOAc, MeCOOH or EtCOOH, (ii) SnCl$_2$, POCl$_3$, C$_5$H$_5$N, (iii) LiPPh$_2$, (iv) MeI.

$$-CH(OH)CH(OH) \xleftarrow{\text{(i)}} -CH=CH- \xrightarrow{\text{(ii)}} -CH(OH)CH(OH)-$$

threo-diol	*cis*-alkene	*erythro*-diol
erythro-diol	*trans*-alkene	*threo*-diol

Figure 7.16 Conversion of alkenes to *vic*-diols. Reagents: (i) *trans*-hydroxylation by I$_2$, AgOCOPh (anhydrous) or epoxidation followed by acid-catalysed hydrolysis, (ii) *cis*-hydroxylation by dilute alkaline KMnO$_4$; I$_2$, AgOAc, AcOH (moist); OsO$_4$; or Oso$_4$ in the presence of an oxidizing agent (metal chlorate, H$_2$O$_2$-Bu^{t}OH, Bu^{t}OOH, or *N*-methylmorpholine *N*-oxide).

polyurethanes from epoxidized oils. The salts of some epoxy acids show rust inhibiting properties in water-based cutting fluids.

7.5.2 Hydroxylation

An alkene can be converted to a diol by reagents which effect *cis* or *trans* addition and the diols have *threo* or *erythro* configuration. The relation between these is set out in Figure 7.16.

Oleic acid yields two racemic diols: the *threo* (mp 95°C) and *erythro* (mp 131°C) isomers. There are eight 9,10,12,13-tetrahydroxystearic acids from the Δ9,12-diene acids and 32 9,10,12,13,15,16-hexahydroxystearic acids from the Δ9,12,15 triene acids.

vic-diols are cleaved by periodic acid or by permanganate to give aldehydic or acidic fragments, respectively, and the diols can be converted back to alkenes stereospecifically by reactions such as those set out in Figure 7.17.

7.5.3 Oxidative fission

Oxidative cleavage of double bonds followed by recognition of the fragments has been an important reaction in the history of fatty acid chemistry because, until recent times, it was the best method of determining double bond position. It is still used occasionally for this purpose even though spectroscopic methods (especially mass spectrometry and ^{13}C NMR spectroscopy – see sections 6.5–6.7) now provide the normal

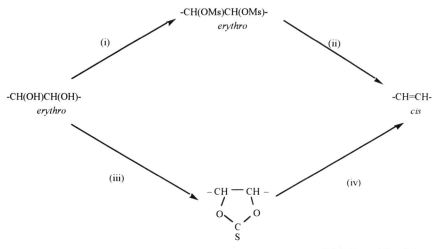

Figure 7.17 Stereospecific conversion of diols to alkenes. Reagents: (i) MsCl, pyridine (Ms = methanesulphonyl); (ii) NaI, Zn, DMF; (iii) thiocarbonyl di-imidazole; (iv) P(OMe)$_3$. (Gunstone and Jacobsburg (1972).

approach to this problem. There remains an interest in this reaction as a way of using fatty acids as a source of useful compounds. Two procedures of oxidative fission have been widely used: von Rudloff oxidation and ozonolysis.

The problem of over-oxidation encountered when olefinic compounds were oxidized with permanganate is overcome in the von Rudloff procedure which employs a mixture of permanganate and periodate in a 1:39 ratio. The periodate regenerates the permanganate as it is used up and there is never a high concentration of oxidizing agent. Acids are oxidized in aqueous alkaline solution and esters in aqueous tert-butanol for 6–12 hours at room temperature.

$$R^1CH{=}CHR^2 \xrightarrow[\text{KIO}_4]{\text{KMnO}_4} R^1COOH + R^2COOH$$

(reaction occurs *via* diols, ketols, and short-chain aldehydes)

Ozonolysis is more often used for both analytical and preparative purposes. Reaction occurs *via* a molozonide and the Criegee zwitterion. This latter reacts with aldehyde to give an ozonide or with alcohols or acids used as solvents to give hydroperoxide derivatives (Figure 7.18).

Unsymmetrical alkenes give symmetrical or unsymmetrical ozonides each of which exists in *cis* and *trans* forms – a total of six compounds (Figure 7.19). This has been demonstrated for a number of compounds including methyl oleate.

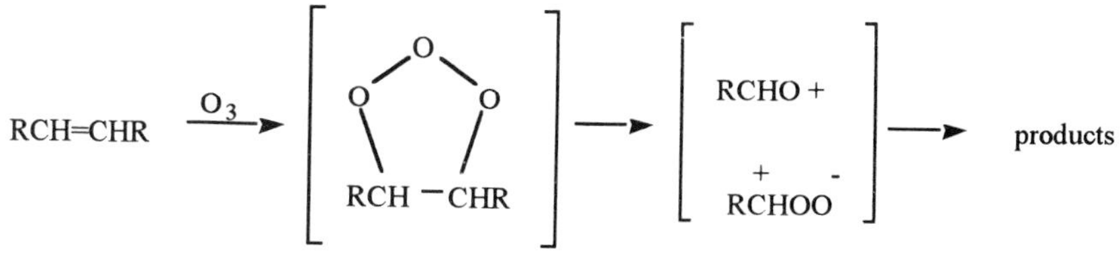

$$RCH=CHR \xrightarrow{O_3} \left[\begin{array}{c} \text{molozonide structure} \end{array} \right] \longrightarrow \left[\begin{array}{c} RCHO + \\ \\ + \quad - \\ RCHOO \end{array} \right] \longrightarrow \text{products}$$

molozonide   zwitterion

reaction products of the zwitterion:

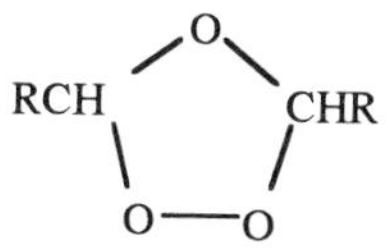

ozonide
(1,2,4-trioxolane)
(from RCHO)

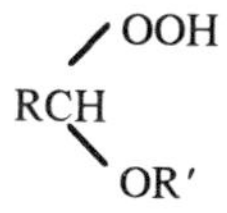

alkoxyhydroperoxide

(from R′OH)

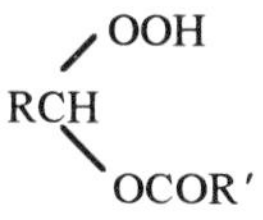

acyloxyhydroperoxide   and   diacyloxyperoxide

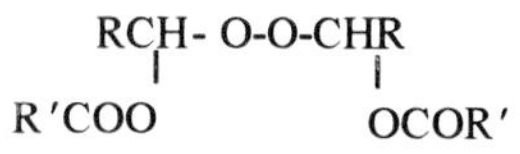

(from R′COOH)

Figure 7.18

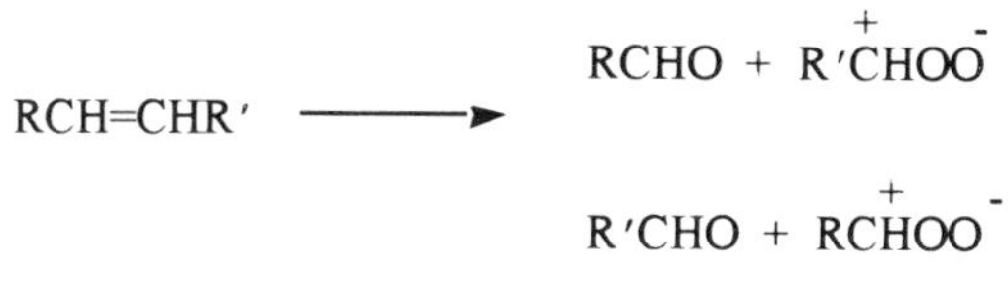

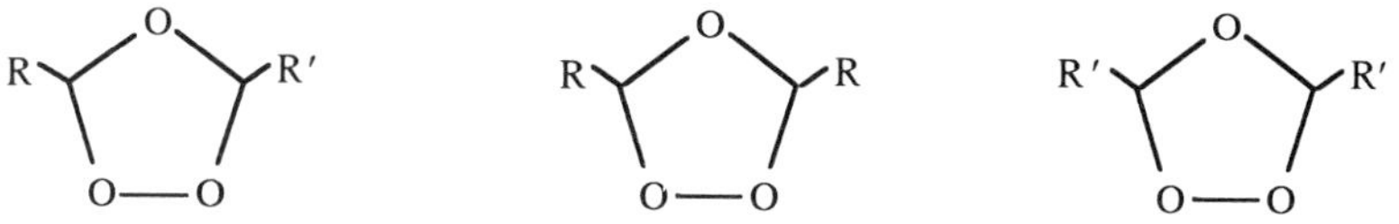

$$RCH=CHR' \longrightarrow \begin{array}{c} + \quad - \\ RCHO + R'CHOO \\ \\ + \quad - \\ R'CHO + RCHOO \end{array}$$

unsymmetrical   symmetrical

c/t   *c/t*   *c/t*

Figure 7.19

The fission products which result from the ozonide or the related hydro-peroxides obtained by reaction with several reagents are more important. These may be alcohols, aldehydes, acids, esters, acetals, or amines with the appropriate reagent (Table 7.12). A monoene acid or ester will give two products, one monofunctional and the other difunctional. For example, oxidative ozonolysis of oleic acids gives a C_9 monobasic acid (nonanoic, pelargonic) and a C_9 dibasic acid (nonanedioic, azelaic).

Ozonolysis of readily available monoene acids is carried out on an industrial scale. The dibasic acid is the more valuable product although uses may also be found for the monobasic acid. Oleic, undecenoic, and erucic acids yield the C_9 (azelaic), C_{10} (sebacic), and C_{13} (brassylic) dibasic acids, respectively. As an example ozonolysis of oleic acid is carried out in a solution of pelargonic (nonanoic) acid at 25–45° which is then heated to 100°C in the presence of a manganese salt.

Industrial ozonolysis presents some difficulties and alternative oxidation procedures are being examined. Two have been reported though there is no evidence that they are being used on a large scale. One employs oxygen in the presence of aldehydes as oxidizing agent and proceeds *via* an epoxide. The second is oxidation with sodium hypochlorite in a three-stage process involving emulsification of oleic acid in water, non-catalytic oxidation with sodium hypochlorite and alkali, and cleavage of the diol with sodium hypochlorite and ruthenuim chloride ($RuCl_3$) as catalyst.

7.6   Halogenation

(a)   *Halogen addition.*   The addition reaction occurring between halogens and olefins is the basis of the classical method for measuring unsaturation by the iodine value. The most common procedure involves reaction of the sample under investigation with an excess of Wijs' reagent (iodine monochloride in acetic acid) for 30–60 min.

Halogenation is a *trans* addition and details of some products are given in Table 7.13. *cis*-olefinic acids give *threo*-dihalides and the *trans*-acids give the *erythro*-isomers. Dienes and trienes furnish several stereoisomeric products.

Table 7.12 Reagents for the breakdown of ozonides to short-chain products

Product type	Reagents
alcohols	$LiAlH_4$, $NaBH_4$, H_2-Ni, H_2-Pt
aldehydes	Zn-acid, Ph_3P, Me_2S, H_2-Lindlar's catalyst
acids	peroxyacid, Ag_2O
esters/acetals	MeOH-HCl
amines	NH_3-Ni

Table 7.13 Halogenated products (mp) from olefinic adids[a]

Acid	Configuration	Chloro (mp)	Bromo (mp)
18:1	9c	37	28.5
	9t	–	29.5–30
18:2	9c12c	123[b]	115[b]
	9t12t	–	78[b]
18:3	9c12c15c	189[b]	183[b]

[a]Monoenes, dienes and trienes give di-, tetra- and hexahalogenated products, respectively.
[b]Accompanied by other isomers not fully identified.

(b) *Modification of the normal halogen addition.* Bromination is normally a two-step process occurring *via* a bromonium ion intermediate. The second step requiring halide anion can be replaced by reaction with a competing nucleophile:

$$-CH=CH- \xrightarrow{Br_2} \overset{Br^+}{-CHCH-} \text{ (bromonium ion)}$$

$$\xrightarrow{Br^-} -CHBrCHBr-$$
$$\xrightarrow{HOH} -CHBrCH(OH)-*$$
$$\xrightarrow{ROH} -CHBrCH(OR)-*$$
$$\xrightarrow{RCOOH} -CHBrCH(OCOR)-*$$

*only one of two possible products is formulated

For example, methoxybromo adducts are easily made by reaction of *t*-butoxy bromide in methanol.

Chlorination of linoleic acid in the presence of water gives tetrachloro-, trichloro hydroxy- and dichloro dihydroxystearic acids. These compounds are formed during the bleaching of wood pulp or related products with chlorine and may also be present in chlorine-bleached flour.

(c) *Iodolactonization.* The formation of iodolactones from appropriate olefinic acids is an interesting example of a modified iodination process. Olefinic acids with $\Delta 4$ or $\Delta 5$ unsaturation react with $KHCO_3$–I_2–KI to give γ- and δ-lactones, respectively. These can be separated from one another by HPLC. The neutral iodolactones are also easily separated from unreacted acidic material.

γ- lactone from Δ4 acid   δ- lactone from Δ5 acid

EPA and AA (Δ5 acids) and DHA (Δ4 acid) react in this way. The γ-lactones are more easily formed and more stable: the δ-lactones are formed less readily and are more reactive. Unsaturated acid can be regenerated by reaction of the iodolactone with trimethylsilyl iodide. The differing reactivities can be manipulated in several ways and the reaction has been used to isolate AA, EPA, and DHA from natural mixtures containing these acids.

(d) *Other halogenation procedures.* Monohalogeno acids are most easily prepared from the corresponding hydroxy acids (e.g. 12-halogeno-stearic acid from 12-hydroxystearic acid) by reaction with triphenyl-phosphine and tetrachloro- or tetrabromomethane or by conversion to methanesulphonate and subsequent reaction with magnesium bromide in ether or magnesium bromide etherate in benzene. Fluorine-containing compounds can be produced from methanesulphonates or 4-toluene-sulphonates by reaction with tetrabutylammonium fluoride or directly from the hydroxy compound by reaction with Et_2NCF_2CHClF or Et_2NSF_3 (DAST).

Carboxylic acids can be converted to 2-halogeno acids or to the *nor*-halides thus:

$$RCH_2Br \xleftarrow{\text{(i)}} RCH_2COOH \xrightarrow{\text{(ii)}} RCHBr\,COOH$$

Reagents: (i) bromine and silver salt, or $HgO + Br_2$, or $Pb(OAc)_4 + I_2$, or thallium salt and bromine; (ii) P, Br_2.

Allylic bromination with NBS or Bu^{t}OBr gives a mixture of isomeric bromo compounds. Heavily brominated jojoba oil was made by allylic bromination, followed first by removal of hydrogen bromide and then by bromination.

$$\text{monoene}\begin{cases} \xrightarrow{\text{NBS}} \text{monobromo} \xrightarrow{\text{-HBr}} \text{diene} \xrightarrow{Br_2} \text{tetrabromide} \\ \xrightarrow{\text{NBS}} \text{dibromo} \xrightarrow{\text{-HBr}} \text{triene} \xrightarrow{Br_2} \text{hexabromide} \end{cases}$$

(e) *Dehalogenation, dihydrohalogenation.* *vic*-dibromo acids are de-brominated by zinc or sodium iodide (or several other reagents). Both effect *trans* elimination, with sodium iodide being the more stereospecific. The resulting alkene has the same configuration as the original olefin and bromination-dehydrobromination can be used as a method of protecting double bonds.

$$-CH=CH- \quad \xrightarrow{\text{Br}_2} \quad -CHBr\ CHBr- \quad \xrightarrow{\text{NaI}} \quad -CH=CH-$$
cis   *threo*   *cis*
trans   *erythro*   *trans*

Double dehydrohalogenation of *vic*-dihalides with a base gives an allene with the *erythro*-dihalides from *trans*-alkenes and acetylenes with the *threo*-dihalides from *cis*-alkenes. This is the basis of the conversion of oleic acid to stearolic acid:

$$-CH=CH- \quad \longrightarrow \quad -CHBr\ CHBr- \quad \longrightarrow \quad -C\equiv C-$$
cis   *threo*

Reaction of 2-halogeno acids with KOH–EtOH gives a mixture of the $\Delta 2t$ alkenoic acid along with 2-hydroxy and 2-ethoxy acids.

7.7 Oxymercuration

Alkenes react with mercury acetate (or other mercury salts) and methanol (or other nucleophiles) to give a mercury-containing adduct in a *trans* addition process. The product can then be converted to a range of non-mercury compounds:

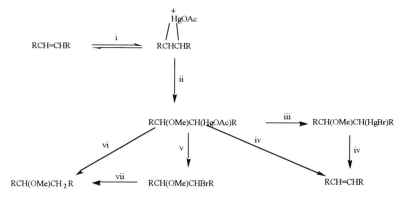

Reagents: (i) Hg(OAc)$_2$; (ii) MeOH, (iii) NaBr, MeOH, (iv) HCl, (v) Br$_2$, (vi) NaBH$_4$, (vii) Bu$_3$SnH. (The mercuribromide and the mercuri-acetate give the same products with reagents (iv), (v), and (vi)).

Polyene esters undergo virtually complete reaction to give products with the appropriate number of mercuriacetate groups. These are separable by TLC, more conveniently as the mercuribromides and can be examined separately by gas chromatography after regeneration of the alkene by reaction with hydrochloric acid. Under slightly different conditions this strategy has also been employed to concentrate polyene acids from natural mixtures in which they occur (e.g. linolenic acid from linseed oil and EPA/ DHA from fish oils).

Reaction of the methoxy mercuriacetates with sodium borohydride gives a mixture of isomeric methoxy acids which can be used to identify double bond position by mass spectrometry (section 6.7.1). The method is less satisfactory with polyenes because of the complexity of the spectra. Better results are obtained by partial reduction, isolation of the mono adducts, and conversion of these to isomeric monomethoxy compounds. In this way linolenate give a mixture of six methoxystearates (9,10,12,13,15, and 16) each of which can be more easily identified by mass spectrometry.

Other substituents can be introduced into the alkanoic acid chain by using other nucleophilic reagents in place of methanol. Thus oleate yields a mixture of the 9- and 10-substituted stearates. Typical examples include OEt (EtOH), OAc (AcOH), OH (H_2O), NHAc (CH_3CN), and OOBu^t (Bu^{t}OOH) which are obtained with the reagents indicated in parenthesis.

Oxymercuration of acetylenic compounds is not reversible and yields oxo or hydroxy compounds depending on the experimental procedures employed.

7.8 Stereomutation

Natural unsaturated acids are almost entirely *cis*-isomers (sections 1.4 and 1.5) and their conversion to *trans* compounds (stereomutation) has always been of interest. This change may occur during processing or it may be promoted with appropriate reagents. *Cis*- and *trans*-isomers have different physical properties (especially melting behaviour) and different nutritional properties.

Isomerization, particularly of linolenic acid, occurs during high tempera-ture steam deodorization (about 250°C) of oils containing this acid such as soybean oil and rapeseed oil. As much as 30% of the all-*cis* triene may be changed – mainly to the *cct* and *tcc* isomers. Isomerization of linoleic acid occurs to a lesser extent being some 13–14 times less reactive than the triene acid. *Trans*-isomers are also formed during partial hydrogenation (section 7.1.1) and such compounds may represent an appreciable proportion of spreading and baking fats. *Trans*-acids in ruminant fats result from biohydrogenation of dietary linoleic and linolenic acids in the rumen. The conversion of oleic acid to its nitrile by reaction with ammonia and a catalyst at 280–360°C (section 8.7.2) is accompanied by some stereomutation of the *cis* double bond.

Stereoisomers can be interconverted by sequences of stereospecific reactions which together represent a stereomutation. Several such procedures exist, many of them involving epoxides or dihydroxy acids. A typical process is formulated below:

$$-CH=CH-\ \xrightarrow{(i)}\ \overset{\displaystyle O}{\overset{\displaystyle /\backslash}{-CHCH-}}\ \xrightarrow{(ii)}\ -CH(OH)CH(PPh_2)\ \xrightarrow{(iii)}\ -CH=CH-$$

$$\textit{cis} \qquad\qquad \textit{cis} \qquad\qquad \textit{threo} \qquad\qquad \textit{trans}$$

Reagents: (i) $ArCO_3H$, a *cis*-addition, (ii) $LiPPh_2$, reaction occurring with inversion at one centre, (iii) MeI, a *cis*-elimination.

More commonly the *cis*-acid (or ester) is treated with a reagent which establishes the *cis/trans* equilibrium. For a monoene this is between 3:1 and 4:1 in favour of the *trans*-isomer. This is higher melting and less soluble than the *cis*-isomer and can usually be isolated by crystallization or by silver ion chromatography (section 5.2.4c). Polyenes such as linoleic and linolenic acids give mixtures of all the possible *cis*- and *trans*-isomers (4 and 8, respectively) with *trans* unsaturation predominating. Conjugated polyene acids isomerize more readily and give a higher proportion of the all-*trans* isomer. In this case the change is simply effected with iodine and light.

The reagent selected should promote stereomutation without double bond movement or hydrogen-transfer. Reaction with selenium at 190–200°C, much used in early studies, is now seldom used because some double bond migration occurs. The more commonly used reagents include sodium nitrite and nitric acid (a source of NO_2, effective at 40°C in a few minutes) and sulphur-containing compounds such as 3-mercaptopropionic acid, 2-mercaptoethanoic acid, 2-mercaptoethylamine, thiophenol, aryl sulphonic acids, and toluenesulphinic acid. Reaction proceeds, presumably, through a reversible addition process which leads eventually to the equilibrium mixture of *cis*- and *trans*-isomers.

$$-CH=CH-\ \rightleftharpoons\ \text{intermediate}\ \rightleftharpoons\ -CH=CH-$$

$$\textit{cis} \qquad\qquad \text{species} \qquad\qquad \textit{trans}$$

The proportion of *trans*-isomer in a mixture can be measured by gas chromatography, infrared spectroscopy, infrared attenuated reflectance spectroscopy, ^{13}C NMR spectroscopy, or preparative Ag^+ thin-layer chromatography.

7.9 Metathesis

The catalysed olefinic metathesis reaction is much used in the petrochemical industry. It has been studied with unsaturated fatty acids (as methyl or glycerol ester) and alcohols (as acetate or trimethylsilyl ether) but finds no commercial application as yet.

The equilibrium process summarized in the equation:

$$2R^1CH{=}CHR^2 \rightleftharpoons R^1CH{=}CHR^1 + R^2CH{=}CHR^2$$

requires a catalyst which may be homogeneous (WCl_6 along with $EtAlCl_2$ or $SnMe_4$ or $SnBu_4$ or BEt_3) or heterogeneous (oxides or carbonyls of molybdenum or tungsten on alumina and/or silica). A catalyst consisting of B_2O_3–Re_2O_7 and Al_2O_3/SiO_2 activated with $SnBu_4$ is reported to be very reactive.

Metathesis of methyl oleate will give a hydrocarbon (9-octadecene), an ester (methyl 9-octadecenoate), and a diester (dimethyl 9-octadecenedioate) in each of which the double bond has a mixture of *cis* and *trans* configuration. The dibasic C_{18} acid has been used to prepare the fragrance material civetone. Methyl oleate and 3-hexene will give methyl 9-dodecenoate among other products, while 10-undecenol (as its trimethylsilyl ether) and 4-octene give 10-tetradecenol as the most interesting product. It is thus possible to modify chain length.

Using the B_2O_3–Re_2O_7 catalyst described above C_{18}, C_{20}, and C_{26} dibasic acids have been made in good yield from oleate, 10-undecenoate, and erucate respectively using a two-step process. This is illustrated with one example.

$$CH_3(CH_2)_7CH{=}CH(CH_2)_{11}CO_2Me \xrightarrow[\text{catalyst}]{C_2H_4} CH_2{=}CH(CH_2)_{11}CO_2Me$$

$$\xrightarrow[\text{catalyst}]{} MeO_2C(CH_2)_{11}CH{=}CH(CH_2)_{11}COOMe$$
$$C_{26} \text{ diester}$$

For each of these steps the conversion rate is between 86 and 99% and the isolated yield between 60 and 84%.

7.10 Double bond migration and cyclization

Double bond migration occurs with basic or acidic catalysts. This is particularly obvious with methylene-interrupted polyene acids when double bond migration leads mainly to conjugated systems, e.g.

linoleate 9c12c	$\longrightarrow$	9c11t and 10t12c
linolenate 9c12c15c	$\longrightarrow$	10t12c15c, 9c11t15c, 9c13t15c, 9c12c14t
		and conjugated trienes such as 10t12c14t

Such patterns of conjugated unsaturation are easily detected by ultraviolet spectroscopy and alkali-isomerization was the basis of a method for measuring linoleate and linolenate before the general use of gas chromatography.

Under acidic conditions double bond migration occurs through a series of protonation–deprotonation processes. With perchloric acid, for example, oleic acid gives several 18:1 isomers and the γ-lactone. This last is formed by interaction of the carbocation at C4 with the carboxyl group. Polyphosphoric acid also gives cyclic enones.

As already discussed, reactions such as hydrogenation and oxidation which involve an allylic centre are often accompanied by double bond migration.

Polyene acids, especially those with three or more double bonds, undergo cyclization at elevated temperatures. The process is promoted by alkali (for double bond migration) and by sulphur or iodine (for stereomutation) since particular polyene systems are required as intermediates. Cyclic acids are minor components of tall oil and are probably formed from the Δ5,9,12 triene acids present in this source.

$$\Delta 5,9,12 \longrightarrow \Delta 5,10,12 \equiv$$

e.g. $R = CH_3(CH_2)_4$ $R^1 = (CH_2)_3COOH$

Extended heat treatment (200–275°C) of oils such as sunflower (rich in linoleic acid) and linseed (rich in linolenic acid) give rise to cyclic fatty acid monomers. These are readily separated from non-cyclic acids by urea fractionation. The products are complex mixtures of *cis*- and *trans*-isomers of 1,2-disubstituted cyclohexanes and cyclopentanes with two double bonds if based on linolenic acid and one double bond if based on linoleic acid. After hydrogenation, to simplify the structure for investigation, the following range of compounds have been identified but there are probably further isomers of both the cyclohexane and cyclopentane type.

$m = 6-9, n = 10-m$

cyclized from Cm+2 to Cm+7

cis- and *trans*-isomers

$m = 5-9, n = 11-m$

cyclized from Cm+2 to Cm+6

cis- and *trans*-isomers

The reaction mixture also contains *cis-trans*-isomers of linoleate (Δ9,12) and of the conjugated diene acids (9,11; 10,12; 11,13; and 12,14) along with corresponding isomers of linolenate.

Cyclic compounds of this type have been identified in deep frying fats at levels up to about 0.6%.

7.11 Dimerization (dimer acids, estolides, isostearic acid, Guerbet alcohols and acids)

Dimerization of unsaturated fatty acids occurs in the presence of radical sources, by heating at temperatures between 260 and 400°C, and under the influence of a clay or other cationic catalyst. This last procedure is used on a commercial basis to meet the demand for dimer acids. Typical manufacturing conditions use a montmorillonite clay catalyst (4%) at 230°C for 4–8 hours followed by distillation to produce a dimer concentrate which contains some trimer.

Reaction occurs with both monoene and diene acids. Tall oil, which contains both these acids, is much used to yield a product which is about 80% dimer and 20% trimer. Monoenes give mainly acyclic and monocyclic dimers whilst dienes give mono- and bicyclic dimers (see below). Erucic acid yields a C_{44} dimer.

The dimer acids are used mainly as polyamides. Dimer acids react with diamines (such as ethylene diamine) to give non-reactive polyamides with excellent adhesive properties. They are used for hot-melt adhesives (e.g. in shoe-making) and also in printing inks and in coatings. Reaction with polyamines (such as diethylenetriamines – $H_2N(CH_2)_2NH(CH_2)_2NH_2$) gives reactive polyamides with a free amine group which can act as a curing agent for epoxy resins. Dimer acids are also used in corrosion inhibitors (as imidazoles) and in lubricants (as esters). Heat-treated vegetable oils are used in letterpress inks. Short-path distilled dimer acids of 95–99% purity are used in polymer applications as hydrophobic blocks.

Despite the wide use of dimer acids their chemical structure is still not clearly known. Structures proposed for acyclic, monocyclic, and bicyclic dimers, even if correct, are only part of the story. Hydrogen exchange converts cyclic olefins to cyclohexane and benzene derivatives. Hydroxy acids, estolides, and tetrahydrofurans and tetrahydropyrans have all been recognized as intermediate products.

acyclic dimer

(C_{36})

monocyclic dimer

(C_{36})

bicyclic dimer

(C_{36})

The monomer remaining after dimerization is a complex mixture of saturated and unsaturated and straight- and branched-chain acids. After hydrogenation and separation, a concentrate of branched-chain saturated acids is obtained. This low-melting product, designated isostearic acid, is used in lubricants and cosmetics, usually in esterified form.

Dimerization is usually carried out in the presence of a little water (1–2%). If this is increased (9–10%) a new product (mono-estolide) is produced in useful yield. These are interesting substances with structures similar to that given below which may find useful outlets. Acidic catalysts give higher yields of estolides but of increasing molecular weight with up to ten ester linkages.

$$CH_3(CH_2)_8CH(CH_2)_7COOH$$
$$|$$
$$OCO(CH_2)_7CH=CH(CH_2)_7CH_3$$

typical mono-estolide from oleic acid

Guerbet alcohols provide another type of dimeric molecule. These are made by heating saturated alcohols at 200–300°C with potassium hydroxide or potassium alkoxide and reaction occurs via the corresponding aldehydes which undergo condensation.

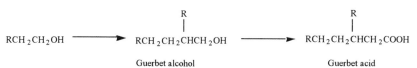

$$RCH_2CH_2OH \longrightarrow RCH_2CH_2\overset{R}{\underset{|}{CH}}CH_2OH \longrightarrow RCH_2CH_2\overset{R}{\underset{|}{CH}}CH_2COOH$$

Guerbet alcohol   Guerbet acid

The product is a branched-chain alcohol which can be oxidized to the corresponding acid. Unlike the dimeric acids these dimers are monofunctional (alcohols or acids). $C_{8–12}$ alcohols give $C_{16–24}$ products and $C_{16–18}$ alcohols give $C_{32–36}$ products. They have depressed melting points and good oxidative stability and are used in cosmetics and as plasticizers and lubricants.

7.12 Other double bond reactions

Double bonds also react with a wide variety of reagents. Several of these have been examined for oleic acid or ester or some other monoene or polyene acids/esters. Some of these are indicated in this section.

(a) *Nitrogen-containing compounds.* The best known nitrogen-containing compounds (other than those associated with the carboxyl group – section 8.7) are the aziridines or epimino compounds and the various amino compounds that can be produced from these. They have a formal resemblance to epoxides and their reaction products and to the corresponding sulphur compounds. The formation of the aziridines and their reactions are summarized in Table 7.14.

(b) *Sulphur-containing compounds.* Some sulphur-containing compounds and their preparation from alkenes are listed in Table 7.15.

Vulcanized vegetable oil is known as factice and is used in rubber compounding to improve quality. Reaction with sulphur dichloride gives white factice, while sulphur itself gives dark factice. The structures of these products are not fully determined.

Table 7.14 Preparation and reactions of aziridines[a]

(i)	Preparation from alkenes		
	reagent	intermediate	reagent[b]
	IN_3	–CHICH(N_3)	$LiAlH_4$
	INCO	–CHICH(NCO)	MeOH, base
	Cl_2NCO_2Et	–CHClCH($NClCO_2Et$)	base
(ii)	Reactions of aziridines		
	reagent	product	
	H^+, MeOH	–CH(NH_2)CH(OMe)–	
	HI	–CH(NH_2)CHI–	
	NaN_3 and NH_4Cl or HN_3	–CH(NH_2)CH(N_3)–[c]	
	RCOOH	$\left\{\begin{array}{l} \text{–CH}(NH_2)\text{CH(OCOR)–} \\ \text{–CH(NHCOR)CH(OH)–} \end{array}\right.$	

$$\overset{\text{NH}}{\triangle}$$
[a]RCH CHR'
[b]These reagents produce the aziridine from the intermediate listed.
[c]Reduced by hydrogen to the diamine.

Table 7.15 Sulphur-containing acids/esters from olefinic compounds

Product	Reagent
$\overset{\text{S}}{\underset{\text{-CHCH-}}{\triangle}}$ [a]	thiourea, thiocyanogen, thiocyanate or other sulphur-containing compounds on the epoxide
–CH_2CH(SH)–[b]	H_2S (radical reaction)
–CH_2CH(SAc)–[b]	CH_3COSH (radical reaction)
–CH_2CH(SCH_2COOH)–	$HSCH_2$COOH (radical reaction)
–CHCH(Cl)– | S | –CHCH(Cl)–	SCl_2

[a]Epithio compounds: 9,10-epithiostearic acid melts at 58°C (*cis*) and 64°C (*trans*).
[b]Also produced from the mesylate by reaction with NaHS or $KSCOCH_3$.

(c) *Cyclopropane and cyclopropene acids.* Cyclopropanes are prepared from alkenes and cylopropenes from alkynes as indicated in the following equations:

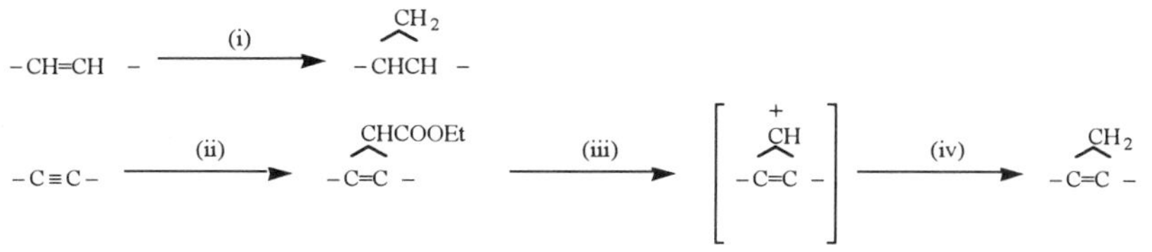

Reagents: (i) CH_2I_2, Cu/Zn (Simmonds-Smith reaction); (ii) $N_2CHCOOEt$, Cu bronze; (iii) FSO_3H or $ClSO_3H$; (iv) $NaBH_4$.

(d) *Addition reactions.* Oleic acid, ester, or alcohol undergo addition reactions with a wide range of reagents. These are mainly acid-catalysed reactions occurring via a carbocation though some are radical processes. The products are usually mixtures because reaction can occur at either olefinic carbon atom and at other carbon atoms in the chain because of the reversible nature of the protonation–deprotonation equilibrium. Some typical reactions of this type are listed in Table 7.16. None of them is exploited on a commercial scale.

Table 7.16 Addition reactions of methyl oleate and related compounds

Reagent	Catalyst	Product[a]	Reaction name
benzene (etc).	$AlCl_3$	$-CH_2CHPh-$	Friedel–Craft
CH_3COCl[b]	$AlCl_3$	$-CH_2CH(COCH_3)-$	Friedel–Craft
syn gas (H_2, CO)	$Co(CO)_8$	$-CH_2CH(CHO)-$[c]	hydroformylation (oxo)
CH_2O	Me_2AlCl or $EtAlCl_2$	$-CH=CHCH(CH_2OH)-$	–
RCN	H_2SO_4	$-CH_2CH(NHCOR)-$	Ritter
CO, H_2O	H_2SO_4	$-CH_2CH(COOH)-$	Koch
ROH (primary)	H_2SO_4	$-CH_2CH(OR)-$	–
CH_3COCH_3	Mn^{II} acetate	$-CH(Z)CH=CH-$ or $-CH(Z)CH_2CH_2-$[d]	–
R_FI[e]	(i) $Sn^{II}Cl$, AgOAc (ii) Bu_3SnH	$-CH(R_F)CH_2-$	–

[a]The product mixture is often less simple than indicated here with the new substituent not confined to the olefinic carbon atoms.
[b]Similar reaction with succinic anhydride.
[c]The formyl function is easily converted to CH_2OH (reduction), COOH (oxidation), $CH(OR)_2$ (acetal), or other derivatives.
[d]$Z = CH_2COCH_3$, the olefinic product is produced in the presence of Cu^{II} and the saturated compound in the presence of a hydrogen donor. Acetone can be replaced by acetic acid or malonic acid.
[e]R_F = perfluoroalkyl such as C_4F_9, C_6F_{13}, or C_8F_{17}.

Bibliography

Allen, J.C. and R.J. Hamilton (eds) (1994) *Rancidity in Foods*, 3rd edn, Chapman and Hall, London.

Boelhouwer, C. and J.C. Mol (1985) Metathesis reaction of fatty acid esters, *Prog. Lipid Res.*, **24**, 243–267.

Cascade Biochem Ltd. (1994) Catalogue, Reading (eicosanoid cascade).

Chan, H.W.S. (ed.) (1987) *Autoxidation of Unsaturated Lipids*, Academic Press, London.

Coppen, P.P. (1990) Antioxidants in food use, *Lipid Technol.*, **2**, 95–99.

Emken, E.A. and H.J. Dutton (eds) (1979) *Geometrical and Positional Fatty Acid Isomers*, American Oil Chemists' Society, Champaign, USA.

Frankel, E.N. (1980) Lipid oxidation, *Prog. Lipid Res.*, **19**, 1–22.

Frankel, E.N. (1982) Volatile lipid oxidation products, *Prog. Lipid Res.*, **22**, 1–33.

Frankel, E.N. (1985) Chemistry of free radical and singlet oxidation of lipids, *Prog. Lipid Res.*, **23**, 197–221.

Frankel, E.N. (1991) Recent advances in lipid oxidation, *J. Sci. Food Agric.*, **54**, 495–511.

Gunstone, F.D. and F.R Jacobsburg (1972) *Chem. Phys. Lipids*, **9**, 112–122.

Gunstone, F.D. *et al.* (eds) (1994) *The Lipid Handbook*, 2nd edn, Chapman and Hall, London (all sections of this chapter).

Hudson, B.J.F. (ed.) (1990) *Food Antioxidants*, Elsevier Applied Science, London.

Johnson, R.W. and E. Fritz (eds) (1988) *Fatty Acids in Industry*, Dekker, New York. (hydrogenation, oxidation, dimerization)

Loliger, J. (1991) Natural antioxidants, *Lipid Technol.*, **3**, 58–61.

Maerker, G. (1987) Cholesterol autoxidation – current status, *J. Amer. Oil Chem. Soc.,* **64**, 388–392.

Patterson, H.B.W. (1989) *Handling and Storage of Oilseeds, Oils, Fats and Meal*, Elsevier, London, pp. 1–86 (chapter on oxidative deterioration).

Patterson, M.B.W. (1994) *Hydrogenation of Fats and Oils: Theory and Practice*, AOCS Press, Champaign, USA.

Pryde, E.H. (ed.) (1979) *Fatty Acids*, American Oil Chemists' Society, Champaign, USA.

Sehedio, J.L. (1993) Mercury adduct formation in the analysis of lipids, in *Advances in Lipid Methodology – Two* (ed. W.W. Christie) The Oily Press, Dundee, pp. 139–155.

Sebedio, J.L. and A. Grandgirard (1989) Cyclic fatty acids: natural sources, formation during heat treatment, synthesis and biological properties, *Prog. Lipid Res.*, **28**, 303–336.

Turner, G.P.A. (1988) *Introduction to Paint Chemistry and Principles of Paint Technology*, Chapman and Hall, London.

Wolff, R.L. (1993) Further studies on artificial geometrical isomers of α-linolenic acid in linolenic acid-containing oils, *J. Amer. Oil Chem. Soc.*, **70**, 219–224.

8 Reactions of the carboxyl group

8.1 Introduction

As stated earlier (section 3.1) about 14% of the world production of oils and fats is used by the oleochemical industry and 90% of this is used for the production of soap and other surface-active compounds. This results from the fact that long-chain compounds based on fatty acids are amphiphilic in character. Each molecule contains an aliphatic chain which is lipophilic/hydrophobic and a polar head group which is hydrophilic/lipophobic and the operation of these opposing properties gives these molecules their characteristic surface-active behaviour. Aspects of this are discussed in section 10.6. Changing the nature of the polar head group, starting most often with glycerol triesters, is therefore of great importance and most oleochemical activity at an industrial level is based on reactions of the carboxyl group. Changes in the alkyl chain depend mainly on the choice of starting material between three major types based on C_{12} (lauric oils), $C_{16/18}$ (palm, tallow, etc.), and C_{22} (erucic oils) chain lengths.

Surface-active compounds are of three main types depending on the polarity of the head group. Anionic surfactants have a negative charge and include salts of sulphates ($ROSO_2OH$), sulphonates (RSO_2OH), and carboxylates ($RCOOH$): nonionic surfactants, although not charged, are polar by virtue of the presence of several oxygen-containing functions and include monoacylglycerols, partially acylated or alkylated carbohydrates, and ethylene oxide derivatives of OH or NH compounds: cationic surfactants have a positive charge and are mainly quaternary ammonium compounds such as $R\overset{+}{N}Me_3\ \bar{X}$.

The main interactions of glycerol esters, methyl and other alkyl esters, alcohols, and nitrogen-containing compounds are discussed in this chapter and are summarized in Figure 8.1.

There is considerable interest in the use of enzymic catalysts for these processes since they offer a number of potential advantages, although, for the most part, these have yet to be realized on an industrial scale. Since enzymic processes occur at temperatures and pressures not far from atmospheric they may require lower capital costs for plant construction and lower operating costs than the conventional chemical procedures. However, since such reactions are slower, they require more equipment for an equivalent throughput. Mainly because of the lower reaction temperatures the products are often cleaner and can therefore be obtained at required

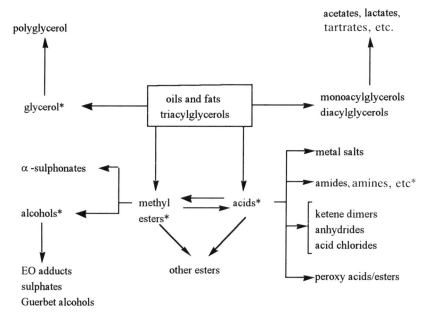

Figure 8.1 Interrelationships between triacylglycerols and their derivatives. EO = ethylene oxide. *major primary oleochemicals.

levels of purity with less waste. This means higher yields and less unwanted material for disposal. Enzymic reactions offer possibilities of greater selectivity and can yield products which cannot be made by non-selective chemical processes. It is this last factor which generally provides the driving force for many research investigations.

8.2 Hydrolysis

Lipid hydrolysis is usually carried out in the laboratory by refluxing oils/ fats with aqueous ethanolic alkali. Acidification of the hydrolysate liberates fatty acids which can be extracted with organic solvent such as ether or hexane. Glycerol remains in the aqueous phase but non-acidic organic compounds (unsaponifiable material) such as hydrocarbons, long-chain alcohols, sterols, isoprenoids, and glycerol ethers accompany the acids into the organic extract. If necessary, the unsaponifiable material can be extracted from the alkaline hydrolysate before acidification (section 5.1.4). The hydrolysis can be carried out quantitatively to determine saponification value or saponification equivalent (section 5.1.4).

Since there is often confusion about the weight relationship between fat, fatty acids, and glycerol it is useful to note that the hydrolysis of glycerol trioleate (100 g) requires water (6.1 g) and will give oleic acid (95.7 g) and glycerol (10.4 g).

The production of soaps by alkaline hydrolysis (saponification) is usually carried out at around 100°C with glycerol being recovered as a second commercial product. The sodium and potassium salts are used as soaps. Other metal salts are used to promote polymerization of drying oils, in the manufacture of greases and lubricants, as an ingredient in plastics formulations, and in animal feeds.

Fats can also be hydrolysed to free acids by water itself in what is probably a homogeneous reaction between fat and water dissolved in the oil phase (fat splitting). This is usually effected in a continuous, high-pressure, uncatalysed, counter-current process at 20–60 bar and 250°C. Lower temperatures are recommended for highly unsaturated oils. Under these high-temperature conditions the products become discoloured and both the fatty acids and the glycerol may be subsequently distilled.

Enzymic hydrolysis of glycerol esters is important in several ways. Lipases, present in seeds and used by the seed to mobilize its energy reserves, promote hydrolysis and most extracted oil contains low levels of free acid (1–5%) accompanied by glycerol di- and mono-esters. The free acids are removed during refining but this is accompanied by loss of oil. Lipolysis can be minimized by inactivation of the lipases before extraction. Enzymic deacylation is the basis of some valuable analytical techniques (section 5.2.6) and is an important part of fat digestion (section 9.2).

Attempts are being made to develop lipolysis on a commercial scale as an alternative to fat splitting. This has been achieved with enzymes such as *Rhizomucor miehei* and *Candida rugosa*. Reaction takes about 20 hours (or longer) at room temperature or 6 hours at 45°C. Under these conditions the products are purer than those obtained under the harsher conditions of fat splitting and can therefore be recovered in good quality with less waste product. The enzymic process will also have lower energy requirements.

Some enzymes effect deacylation only of the fatty acids attached to the α-positions (*sn*-1 and *sn*-3) but others remove all three acyl groups. Another type of specificity depends on whether or not the acyl chains have a double bond close to the carboxyl group. Acids with unsaturation at Δ4, Δ5, or Δ6 react less readily both in hydrolysis and esterification and attempts to exploit this are described in section 8.3.1.

8.3 Esterification (see also sections 2.9.5–2.9.7)

As summarized in Table 8.1 esters can be made from acids (or related acyl derivatives) or from existing esters by exchange of alkyl or acyl groups.

Table 8.1 Routes for the preparation of esters[a]

Reactants		Reaction type
carboxylic acid[b]	alcohol	esterification
ester[c]	alcohol	alcoholysis[d]
ester[c]	acid	acidolysis
ester[c]	ester	interesterification[e]

[a]All these reactions require a catalyst which is usually acidic or basic or enzymic.
[b]Can be replaced by acid chloride or anhydride – a catalyst is not then normally required.
[c]These reactions, requiring at least one ester, are known collectively as transesterifications.
[d]Important alcoholysis reactions include methanolysis and glycerolysis to give methyl and glycerol esters, respectively.
[e]Also called ester–ester interchange.

8.3.1   Esterification reactions between acids (and related acyl derivatives) and alcohols

The preparation of methyl esters is an important part of the gas chromatographic investigation of free acids since these are the derivatives normally used for this purpose. They can be made from fatty acids by reaction with excess of methanol at 50°C or at reflux temperature in the presence of an acidic catalyst such as sulphuric acid (1–2%), hydrogen chloride (5%, conveniently generated by reaction of excess methanol with acetyl chloride), or boron trifluoride (12–14%), although it is usually more convenient to make methyl esters directly from triacylglycerols by methanolysis (section 8.3.3). When acidic conditions have to be avoided, as with cyclopropane, cyclopropene, epoxy, or allylic hydroxy groups, then diazomethane may be used for methylation.

When it is not appropriate to use excess of the alcohol component, as in the synthesis of glycerol esters, equivalent or near-equivalent quantities of alcohol are employed and acylation is carried out with acid chloride, acid anhydride, or free acid and an appropriate catalyst (section 2.9.2).

Yet another procedure involves reaction of the silver, sodium, potassium, calcium, or tetramethylammonium salt with an alkyl halide.

Using Lipozyme as catalyst (trade name for a preparation of the lipase from *Mucor miehei*) stearic acid has been converted to its ethyl ester in quantities up to 2 kg and canola oil deodorizer distillate to its methyl esters in high yield at about 50°C. Reaction of glycerol with acid (or with methyl, vinyl, or glycerol ester) gives 1-monoacyl or 1,3-diacylglycerol esters depending on the reaction conditions employed. Reacted with a concentrate of the n-3 polyene acids from cod liver oil, glycerol gives a mixture of mono and diacylglycerols rich in EPA and DHA.

The reduced ability of some lipases to catalyse reactions of acids with a double bond close to the carboxyl group such as DHA, EPA, and GLA, with unsaturation starting at positions 4, 5, and 6, respectively, is illustrated in the following enzymic esterification reactions. When the mixed acids of evening primrose oil containing 9.5% of γ-linolenic acid were reacted with butanol and Lipozyme at 30°C the less reactive Δ6,9,12 triene acid remained mainly unesterified. After 4 and 6 hours the free acids contained 82.8% (9.3% yield) and 84.7% (7.2% yield) of γ-linolenic acid, respectively. Similarly with the acids of cod liver oil the level of DHA was raised from 9.4% to 39.5% (10.3% yield after 4 h), 45.9% (7.4% yield after 6 h), and 54.3% (2.0% yield after 16 h).

8.3.2   Acidolysis: reaction between esters and acids

This process involves the interaction of an ester and a carboxylic acid leading to an exchange of acyl groups in the presence of a catalyst at about 150°C. Catalysts that have been employed include sulphuric acid, metal oxides (zinc, calcium, magnesium, aluminium) or mercuric sulphate. For example, reaction between vegetable oils and lauric acid results in the random replacement of C_{16} and C_{18} acids by the C_{12} acid. Reaction between rapeseed oil and lauric acid (3:1 mixture) in the presence of Lipozyme (a 1,3-specific lipase preparation, see section 8.3.4) at 50–70°C leads to the incorporation of about 20% of lauric acid into the mixed triacylglycerols.

8.3.3   Alcoholysis: reaction between esters and alcohols (methanolysis, glycerolysis)

The catalysed reaction between an ester and an alcohol which leads to an exchange of the alkyl portion of the esters is important, particularly as a method of converting triacylglycerols to methyl esters by reaction with methanol (methanolysis) or to mono and diacylglycerols by reaction with glycerol (glycerolysis).

Methanolysis is carried out on a small scale to convert glycerol esters to methyl esters for chromatographic study and is usually effected with an acidic (sulphuric acid, hydrochloric acid, boron trifluoride) or basic (sodium methoxide) catalyst according to the following reactions:

acidic methanolysis

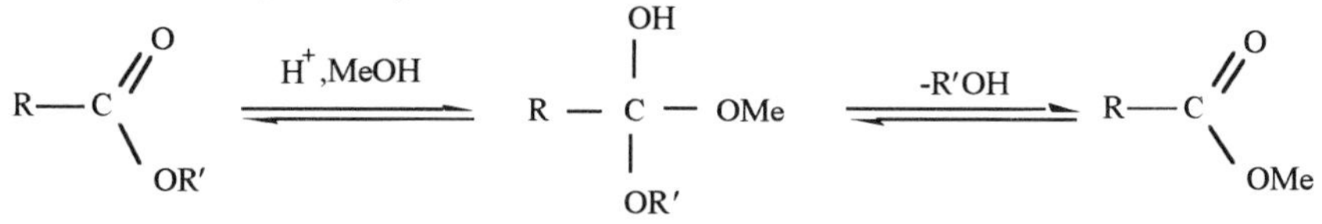

basic methanolysis

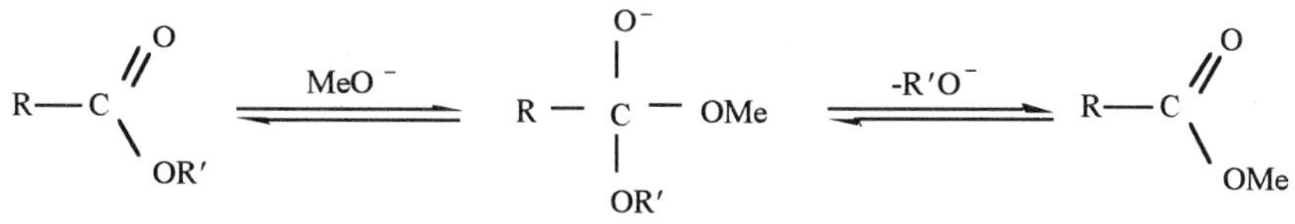

These are equilibrium processes and methanolysis requires a large excess of methanol. The basic process is quicker, being complete in a few minutes when carried out on the milligram scale, but free acid present in the oil will not be esterified and may destroy the catalyst if present at high levels.

Tetramethylguanidine has been recommended as a rapid transesterification catalyst (2 min at 100°C) which will also esterify free acids which may be present (Schuchardt and Lopes, 1988).

Methanolysis is also effected on a large scale to make methyl esters for use as biodiesel or for subsequent reduction to alcohols. In one procedure crude oils with up to 30% of free acids are first reacted with methanol at 80 ± 5°C in the presence of a solid acid catalyst such as a sulphonated ion exchange resin. This converts the free acids to esters and is followed by methanolysis of the glycerol esters using an alkaline catalyst, such as sodium hydroxide, at 70 ± 5°C. Oils low in free acid can be converted directly to methyl esters with an alkaline catalyst. Glycerol is also produced in this reaction and is recovered as a second saleable product. Isopropyl esters are made by reaction of glycerol esters or methyl esters and isopropanol in the presence of an acidic catalyst.

When a triacylglycerol is heated with glycerol and a basic catalyst such as sodium hydroxide or sodium methoxide the following equilibrium is established:

triacylglycerol + glycerol $\rightleftharpoons$ monoacylglycerol + diacylglycerol

The composition of the equilibrium mixture depends on the amount of glycerol dissolved in the lipid phase. This is an important route to mixtures of monoacylglycerols and diacylglycerols and 90–95% concentrates of monoacylglycerols are obtained by molecular distillation. These are used extensively as emulsifiers (section 10.5).

Monoacylglycerols can also be made with enzymic catalysts by acylation of glycerol or by glycerolysis. The yield of monoacylglycerol obtained by enzymic glycerolysis is temperature dependent. The optimum temperature is higher with solid fats and lower with liquid oils. Using *Pseudomonas* sp the yield of monoacylglycerols was around 90% for tallow, palm oil, and palm stearin when reaction was effected at 42°C for 8–16 h followed by 4 days at 5°C. In another report hydrogenated beef tallow and glycerol gave diacylglycerols in 90% yield after reaction at 60°C for 2 h followed by 55°C for 4 h and 48°C for 3 days.

8.3.4 *Interesterification: reaction between esters and esters* (see also section 4.5)

The physical properties of a fat depend on its triacylglycerol composition which is related both to the fatty acids present and to their distribution among the triacylglycerols. Interesterification of a single oil leads to rearrangement of the acyl groups, while interesterification of blends changes both fatty acid composition and triacylglycerol composition. The changes usually involve a shift from non-random towards random distribution of acyl groups. This is most simply demonstrated by showing that after interesterification the fatty acid composition of the 2-mono-acylglycerols remaining after lipolysis (section 5.2.6) is the same as that of the whole oil. The triacylglycerol composition of a fully randomized fat can be calculated directly from the fatty acid composition. Thus the six isomeric triacylglycerols containing acids A ($a\%$), B ($b\%$), and C ($c\%$) will each be present at a level of 100 [$a/100 \times b/100 \times c/100$]% and the proportion of other triacylglycerols can be derived similarly.

Catalysts employed at a 0.2–0.4% level include sodium hydroxide, sodium methoxide, or a sodium–potassium alloy. Reaction occurs at about 80°C over 30–60 min. The oil or fat should be as free as possible of water, free acid, and hydroperoxide since these destroy the catalyst. The actual catalyst is probably a glyceride anion (Figure 8.2). The reaction product, which contains free acid (from NaOH) or methyl ester (from NaOMe) and some diacylglycerols, has to be purified.

When applied to a single oil the redistribution of acyl groups from non-random to random distribution changes triacylglycerol composition and leads to a change in melting behaviour. The melting point of soybean oil is raised from −7 to +6°C, that of cottonseed oil from 10 to 34°C, while for hardened palm kernel oil the melting point is reduced from 46 to 35°C. When applied to a mixture of fats all the acids will be redistributed at random and this provides a method of transferring saturated acids to predominantly unsaturated oils and vice versa. Despite the many claims for complete randomization there is some evidence that the changes are not entirely random.

The preparation of triacylglycerols by reaction of glycerol triacetate with

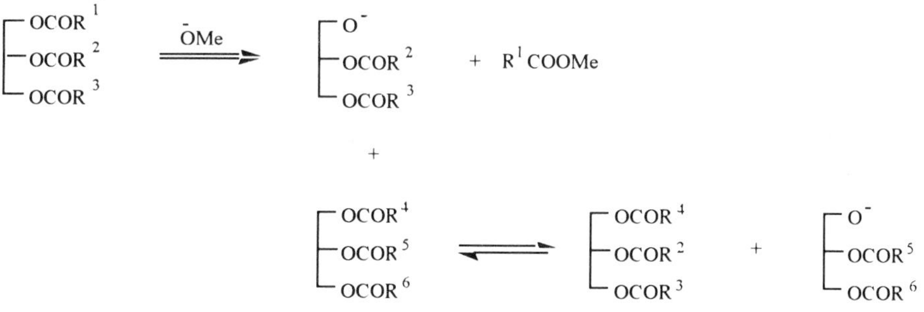

Figure 8.2 Interesterification of triacylglycerols with sodium methoxide as added catalyst. The glyceroxide anion is the true catalyst. The reaction set out above shows the transfer of an acyl group from one triacylglycerol to another. Reaction is not confined to the positions shown but can occur at any acyl group in any intramolecular or intermolecular manner.

the methyl esters of long-chain acids with complete removal of acetate is another example of interesterification. The equilibrium is displaced in the desired direction because methyl acetate is volatile (bp 57°C) and easily removed from the reaction mixture.

Interesterification has been used industrially to improve the physical properties of lard and to produce cocoa butter substitutes from cheaper oils (usually also involving hydrogenation and/or fractionation), fats containing acetic acid, and margarine combining appropriate melting behaviour with a minimum content of *trans*-acids and a maximum level of polyethenoid acids. For example, this last has been achieved by interesterification of soybean oil (80%) and fully hydrogenated soybean oil (20%).

These changes can be further modified if interesterification is carried out at lower temperatures (0–40°C). Under these conditions the fully saturated glycerol esters crystallize from the reaction mixture and thus disturb the equilibrium in the liquid phase. As a consequence the final product contains more S_3 and U_3 and less of the mixed triacylglycerols and this leads to a wider melting range. The reaction is slower at this temperature and may require up to 24 h.

$$U_2S + US_2 \rightleftharpoons U_3 + S_3 \; (U = \text{unsaturated } S = \text{saturated})$$

There is considerable interest in enzymic interesterification and some processes have been commercially exploited. The special value of such catalysts is the additional control of product composition through the differing specificities of the enzymes. The lipase is usually coated onto support material such as kieselguhr, hydroxylapatite, alumina, or phenol-formaldehyde resin and effects interesterification in the presence of a little water at 40–80°C over 16–70 hours. The process can be carried out in a batch or continuous process and reaction is usually of an oil/fat with a second oil/fat, acid, or methyl ester. If the second component is an acid then strictly the reaction is an interesterification *via* acidolysis.

Some enzymes are not regiospecific: reaction then occurs at all three ester sites. With these the product is finally randomized and therefore the same as that obtained more quickly with the ordinary alkaline catalysts. Another group of enzymes react only with acyl groups at the *sn*-1 and *sn*-3 positions and are described as 1,3-specific. Acyl groups at the *sn*-2 position remain unchanged. A third class of enzymes is specific for acids with $\Delta 9$ unsaturation such as oleic, linoleic, and linolenic and only such acids can be removed or inserted into glycerol esters (Table 8.2).

Examples of interesterification and acidolysis are (P = palmitic, O = oleic, St = stearic, U = unsaturated, S = saturated).

1. Sunflower or safflower oil and ethyl stearate, under the influence of an immobilized enzyme from a strain of *Rhizopus niveus*, give a symmetrical triacylglycerol (around 80%) consisting mainly of StOSt along with

Table 8.2 Lipases available for interesterification

Non-specific[a]	1,3-specific[a]	Δ9-specific[a]
Candida cylindricae	Aspergillus niger	Geotrichum candidum
Corynebacterium acnes	Mucor javanicus	
Staphylococcus aureus	Mucor miehei[b]	
	Rhizopus arrhizus	
	Rhizopus delemar	
	Rhizopus niveus	

[a]These terms are explained in the text.
[b]Available as a commercial preparation under the name Lipozyme.

POSt, and POP. The product could be used as a substitute for cocoa butter in chocolate and confectionery products.

2. Vegetable oil and food-grade fatty acids in a solvent-free system with an immobilized lipase from genetically modified *Aspergillus oryzae* give a product which can be used in chocolate and confectionery. An enzymic method of removing unwanted diacylglycerols is also reported.

3. Palm mid-fraction rich in SUS glycerol esters (POP 58%, POSt 13%, and StOSt 2%) yields a product more like cocoa butter after acidolysis with stearic acid (POP 9%, POSt 32%, and StOSt 13%).

4. Acidolysis of high-oleic sunflower oil with stearic acid in the presence of *Mucor miehei* gives a product from which a fraction rich in StOSt (79%) can be isolated.

8.3.5 Structured lipids

Structured lipids are glycerol or sugar esters made to have particular composition and with special physical or nutritional properties. Their preparation requires appropriate acylation procedures which illustrate the methods discussed in the previous sections.

(i) *Medium chain triglycerides (MCT).* Distillation of the acids from coconut oil or palm kernel oil gives an early fraction consisting of octanoic (around 60%) and decanoic acids (about 40%) with less than 1% of each of the C_6 and C_{12} homologues. This mixture is converted to triacylglycerols by catalysed reaction with glycerol. Excess of the acid mixture is usually employed since it is important that the product contains no monoacyl-glycerols as these have a bitter taste. Equally, after refining, the level of free acids should be very low because of the goat-like odour of the C_8 and C_{10} acids.

(ii) *Caprenin.* This is the trade name of a reduced calorie fat (about 5 kcal/g) with functional properties similar to those of cocoa butter and usable in soft candy and in coatings for nuts, fruits, and biscuits. It is a mixture of triacylglycerols made from an equimolar mixture of C_8 and C_{10} acids from lauric oils and of behenic acid (22:0) obtained by hydrogenation of erucic acid. In one method of preparation glycerol behenates (mono, di, and triesters) are esterified with decanoic acid and the resulting triacyl-glycerol is interesterified (NaOH, 80) with medium chain triglycerides with appropriate levels of the C_8 and C_{10} acids. The product is mainly glycerol triesters with one behenic chain (around 92%) and the major triacylglycerols are C_{38} (about 22%), C_{40} (around 48%), and C_{42} (roughly 24%). The long-chain acid is poorly absorbed and this accounts for the low caloric value of this product.

(iii) *Olestra.* This is the trade name of a sucrose polyester with 6–8 acyl groups per sucrose unit made from sucrose and the fatty acids of soybean, corn, cottonseed, or sunflower oil. Olestra is not metabolized and has a virtually zero calorie value.

This sugar ester is produced by reaction between high quality sucrose and high quality methyl esters in the presence of a basic catalyst (alkali carbonate) and a sodium or potassium soap to aid contact between the reactants. The product is finally bleached and deodorized. It can be used as a frying oil or as a replacement for fat in ice-cream, margarine, cheese, and baked goods. It is non-toxic and non-carcinogenic but permission has not yet been granted for its use in food.

(iv) *Betapol.* This is used in infant milk formulae and is unusual in that, like human milk fat, it contains palmitic acid in the *sn*-2 position. It is made by interesterification of tripalmitin with high-oleic sunflower oil or soybean oil fatty acids in the presence of a 1,3-specific enzyme and consists mainly of UPU glycerol esters (P = palmitic, U = unsaturated, oleic or linoleic).

8.4   Acid chlorides, anhydrides, and ketene dimers

Acid chlorides. The most common laboratory preparation of acid chlorides involves reaction at room temperature over 3–5 days of saturated acids with thionyl chloride ($SOCl_2$) or of unsaturated acids with oxalyl chloride ($ClCOCOCl$). Other reagents which can be used include phosgene ($COCl_2$) and phosphorus compounds such as the trichloride, pentachloride, and oxychloride, or triphenylphosphine and tetrachloro-methane. Phosphorus trichloride or phosgene are most often used for large-scale reactions.

Acid anhydrides. Anhydrides are made from acids or acid chlorides by reaction with acetic anhydride or with DCC as formulated below:

$$RCOOH \xrightleftharpoons[\text{}]{Ac_2O} RCOOCOCH_3 \rightleftharpoons RCOOCOR$$

$$RCOCl \xrightarrow{Ac_2O} RCOOCOR \xleftarrow{DCC} RCOOH$$

(DCC = dicyclohexylcarbodi-imide)

Both the acid chlorides and the anhydrides are effective acylating agents in the syntheses of acylglycerols and phospholipids (section 2.9). With improved catalysts they are often replaced by the carboxylic acids themselves.

Alkyl ketene dimers (AKD).   The ketene dimer from hydrogenated tallow is used for sizing paper. It is made from the acid chloride by slow reaction with water and associates with paper (cellulose) by both physical and chemical forces.

$$RCH_2COCl \longrightarrow \begin{array}{c} RCH=C-O \\ | \quad | \\ RCH-C=O \end{array} \xrightarrow{cell\text{-}OH} cell\text{-}OCOCHRCOCH_2R$$

AKD

8.5   Peroxy acids and esters

Acyl derivatives of hydrogen peroxide include the peroxy acids, their esters (especially t-butyl), and diacyl peroxides:

peroxy acid   t-butyl peroxy ester   diacyl peroxide

Although these compounds have been widely studied and safely used on a large scale they should always be handled with care. In particular, they should not be ground, heated in a closed container, inhaled as vapour, or ingested as dust.

Peroxy acids (formerly peracids).   The most important route to the peroxy acids involves establishment of the acid-catalysed equilibrium between acid (or anhydride) and hydrogen peroxide:

$$RCOOH + H_2O_2 \xrightleftharpoons[\text{}]{H^+} RCO_3H + H_2O$$

Sometimes the carboxylic acid is in excess as with formic and acetic. For acids insoluble in water or hydrogen peroxide the reaction may be carried out in concentrated sulphuric acid or in methanesulphonic acid. These act as solvent and as catalyst. The more important peroxy acids are discussed in section 7.5.1.

In contrast to carboxylic acids which form H-bonded dimers the peroxy acids exist as H-bonded monomers (see structure above). This influences the properties of these acids: acidity is reduced 1000-fold and O–H stretching in the infrared is at 3250 cm^{-1} compared with 3530 cm^{-1} for carboxylic acids.

The peroxy acids are useful oxidizing agents. They convert alkenes to epoxides (section 7.5.1), ketones to esters (Baeyer–Villiger reaction), thioethers to sulphoxides (R_2SO) and sulphones (R_2SO_2), and tert amines to N-oxides (R_3NO).

The C_8 and C_9 monobasic peroxy acids and the C_{12} dibasic peroxy acid are being investigated as laundry bleaches. The perborate normally used is less suitable at lower wash temperatures but the organic peroxy acids are effective at these temperatures.

t-Butyl peroxy esters.   These peroxy esters are produced by reaction of t-butyl hydroperoxide with acid chlorides or with carboxylic acids and a catalyst:

$$Bu^tOOH \xrightarrow[\text{RCOOH}]{\text{RCOCl or}} RCOOOBu^t$$

Above 75° they undergo smooth thermal decomposition to give radicals which can be used to initiate radical reactions:

$$RCOOOCMe_3 \xrightarrow{\text{heat}} \begin{array}{l} RCOO\bullet \longrightarrow R\bullet + CO_2 \\ + \\ Me_3CO\bullet \longrightarrow \bullet CH_3 + CH_3COCH_3 \end{array}$$

Diacyl peroxides.   Made from hydrogen peroxide or its metal derivatives, they decompose between 20 and 100°C and are used as radical initiators:

$$\left.\begin{array}{l} Na_2O_2 + RCOCl \longrightarrow \\[2em] H_2O_2 + RCOOH \xrightarrow[\text{DCC}]{} \end{array}\right] RCOOOCOR \xrightarrow{\text{heat}} 2RCOO\bullet \longrightarrow 2R\bullet + 2CO_2$$

8.6   Alcohols

The production of long-chain alcohols is an important industrial process carried out to the extent of about one million tonnes per annum and growing. The C_{16} and C_{18} alcohols are derived from tallow, palm, or palm stearin. The C_{12} alcohol can be made from lauric oils or from cyclododecene produced from ethylene by the petrochemical industry.

The reduction (hydrogenolysis) can be carried out with glycerol esters, methyl esters, or free acids. The first of these is seldom used because glycerol is lost and this is an important element in the economic viability of the process. Reduction of methyl esters is the favoured route. The process is a catalytic reduction, normally occurring at high temperature and pressure (250–300°C, 200–300 bar). With copper chromite as catalyst double bonds are also reduced but with zinc chromite this does not occur and methyl oleate gives oleyl alcohol. The product may be contaminated with some over-reduced product (hydrocarbon) and it is important to minimize this. A new catalyst which does not contain chromium is reported to be effective at lower temperatures and pressures (190–250°C, up to 70 bar) but this has yet to be fully demonstrated on a commercial scale.

Despite the fact that the ester route is in favour at the present time arguments have been presented for the acid route using some new technologies. This would involve (i) conversion of oil to acids and fractionation of these, some of which may be sold as acids, (ii) preparation of methyl ester from acids using a resin bed as catalyst, and (iii) reduction of esters to alcohols using a fixed bed catalyst (40 bar, 200–250°C, chromium-free catalyst). These combined procedures can be used flexibly to produce acids, methyl esters, and alcohols as required. In the laboratory the alcohols are most easily produced with lithium aluminium hydride.

The long-chain alcohols show very good environmental and toxicological properties and are used extensively as surfactants mainly in the form of the sulphate (anionic), ethoxylate (non-ionic), ether sulphate (anionic), or after conversion to an alkyl polyglucoside.

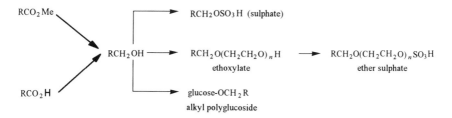

Dimeric alcohols (Guerbet alcohols) are produced from monomeric alcohols by reaction with alkali (section 7.11). These can be oxidized to Guerbet acids and both these branched-chain compounds are used as lubricants and in cosmetics. C_{36} diols can also be produced by reduction of dimer acids.

$$RCH_2CH_2CH(R)CH_2OH \qquad\qquad RCH_2CH_2CH(R)COOH$$

Guerbet alcohol   Guerbet acid

8.7   Nitrogen-containing compounds

8.7.1   Amides

Fat-based amides, amines, and other nitrogen-containing compounds are manufactured at a level of about 300 000 t/a. They are used mainly as surfactants with the long chains derived from coconut (C_{12}), tallow or tall oil (C_{16}, C_{18}), high erucic rape (C_{22}), or from technical grade acids of varying degrees of purity.

Simple amides are made by reaction of ammonia with acids (180–200°C, 3–7 bar, 10–12 h) or esters (220°C, 120 bar, 1 h). The amides based on oleic acid and erucic acid are important compounds incorporated into polyethylene film at a 0.1–0.5% level as anti-slip and anti-block agents. They are also used in printing inks, as lubricant additives, and as mould-release agents.

The more water-soluble ethoxylated amides are made from amides and ethylene oxide at 150–200°C or from esters and diethanolamine at 115–160°C in the presence of base.

$$RCONH_2 + \; n\; \overset{O}{\overset{\displaystyle\triangle}{CH_2\;\;CH_2}} \xrightarrow{\;base\;} RCONH(CH_2CH_2O)_nH$$

$$RCOOCH_3 + HN(CH_2CH_2OH)_2 \xrightarrow{\;base\;} RCON(CH_2CH_2OH)_2$$

Diamides, made from amides and formaldehyde or from acids and ethanediamine, are used as lubricants:

$$2RCONH_2 + CH_2O \longrightarrow RCONHCH_2NHCOR$$

$$2RCOOH + H_2N(CH_2)_2NH_2 \longrightarrow RCONH(CH_2)_2NHCOR$$

8.7.2   Nitriles, amines, and their derivatives

These compounds are interrelated through the sequence shown in Figure 8.3 and have the structures and uses listed in Table 8.3

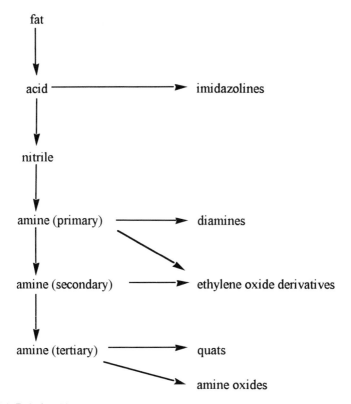

Figure 8.3 Relationships between fats, carboxylic acids, and nitrogen-containing derivatives.

Table 8.3 Typical structures and uses of nitrogen-containing derivatives

Amines	Structures	Uses
primary and secondary	RNH_2 R_2NH	(i) (ii)
ethylene oxide adducts	–	(iii) (iv) (v)
tertiary	R_3N	(i) (iii) (iv) (v) (vi) (vii) (viii)
oxides	R_3NO	(ix) (xiv)
quats	$RNMe_3\bar{C}l$, $R_2\overset{+}{N}Me_2\bar{C}l$	(vi) (x) (xi) (xii) (xiii) (xv)
diamines and their ethylene oxide adducts	$RNH(CH_2)_3NH_2$	(i) (ii) (viii) (x) (xv) (xvi)
imidazolines	see text	(x)
amphoterics	$R\overset{+}{N}H_2(CH_2)_2CO\bar{O}$ $R\overset{+}{N}(Me)_2CH_2CO\bar{O}$	(xii)

(i) flotation agents	(ii) lubricant additives	(iii) emulsifying agents
(iv) foaming agents	(v) corrosion inhibitors	(vi) bactericides
(vii) fungicides	(viii) fuel additives	(ix) cosmetics
(x) fabric softeners	(xi) textile softeners	(xii) drilling muds
(xiii) hair and body care preparations	(xiv) cleaners and detergents	(xv) asphalt treatment
(xvi) water-repellent agents		

Acids are converted to nitriles (probably via amides) by reaction with ammonia (280–360°C with zinc oxide, manganese acetate, bauxite, or cobalt salts as catalyst) and these are used almost entirely to produce primary and secondary amines by hydrogenation in the presence of ammonia and a nickel or cobalt catalyst (120–180°C, 20–40 bar for primary amines, higher temperature and lower pressure for secondary amines). They are converted to tertiary amines ($RNMe_2$, R_2NMe) by catalytic reaction with formaldehyde and to quaternary ammonium compounds or quats by further reaction with alkylating agents such as methyl chloride or sulphate or benzyl chloride. The quats may have 1–3 long alkyl chains, i.e. $R\overset{+}{N}Me_3\bar{C}l$, $R_2\overset{+}{N}Me_2\bar{C}l$, and $R_3\overset{+}{N}Me\bar{C}l$. Tertiary amines can also be made directly from long-chain alcohols and dimethylamine in the presence of thallium sulphate at 360°C. Amine oxides are made from tertiary amines and hydrogen peroxide.

Diamines ($RNH(CH_2)_3NH_2$) are produced from amines or alcohols (RNH_2, ROH) and acrylonitrile ($CH_2=CHCN$). Imidazolines, such as those shown below, are produced from carboxylic acids and appropriate di- and tri-amines.

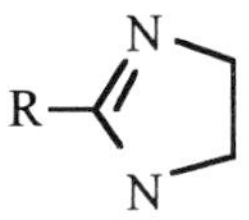

from $H_2N(CH_2)_2NH_2$ 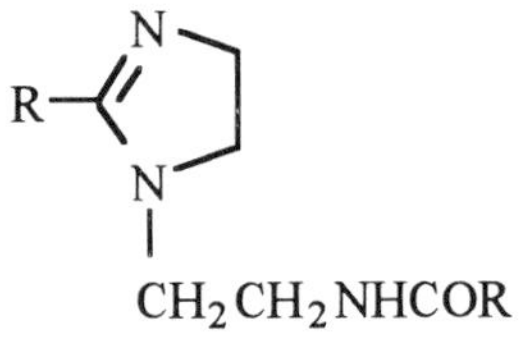   from $H_2N(CH_2)_2NH(CH_2)_2NH_2$

CH_2CH_2NHCOR

8.8 Other reactions of the carboxyl group

The carboxyl group – whether as free acid or ester – raises the reactivity of the adjacent CH_2 group (C2) above that of other methylene groups in long-chain saturated acids. The α-substitution process is sometimes accompanied by decarboxylation:

$$RCH_2COOH \longrightarrow RCHXCOOH \longrightarrow RCH_2X + CO_2$$

These reactions are generally less suitable for unsaturated acids because allylic CH_2 groups are also reactive and the reactions become more complex.

Sulphonation. Saturated acids and esters react with oleum, sulphur trioxide, or chlorosulphonic acid in the presence of chloroform, tetrachloromethane, tetrachloroethene, or liquid sulphur dioxide as solvent.

$$RCH_2COOR' \longrightarrow RCH(SO_3H)COOR'$$

$$R' = H \text{ or } CH_3$$

The products may exist as acid salts (SO_3Na, COOH), neutral salts (SO_3Na, COONa), or esters (SO_3H, $COOCH_3$) and are used in detergent bars (sodium or amine salts) and as viscosity-reducing agents, corrosion inhibitors, and ore-flotation agents. These compounds show high levels of activity, are less toxic in aqueous environments than other surfactants, and are biodegradable in a few days.

α-*anions.* Saturated and monounsaturated acids form dianions with lithium diisopropylamide in hexamethylphosphoric acid. These react with a wide range of compounds to give products such as those listed in Table 8.4.

$$RCH_2COOH \xrightarrow{(Pr^i)_2NLi} \underset{\text{dianion}}{R\overset{-}{C}HCO\overset{-}{O}} \longrightarrow \begin{array}{c}\alpha\text{-substituted} \\ \text{product}\end{array}$$

A range of saturated and unsaturated esters have been converted to succinates

$$RCH_2CO_2Me \xrightarrow[CuBr_2]{(Pr^i)_2NLi} \begin{array}{c} RCHCO_2Me \\ | \\ RCHCO_2Me \end{array}$$

Table 8.4 Products from α-anions

Product	Reagent
RCHDCOOH	D_2O
RCHR'COOH	R'Br
RCH(CH_2OH)COOH $\Big\}$	
RC(=CH_2)COOH	CH_2O
RCH(OOH)COOH	O_2, $-78°C$
RCH(OH)COOH	O_2, $20°C$
RCH(COOH)$_2$	CO_2, $ClCO_2Et$, or $(R'O)_2CO$
RCHXCOOH	X_2(Br_2 or I_2)
RCH$_2$CHO[a]	HCOOH
RCH$_2$NO$_2$[a]	$PrNO_2$
RCH(NH_2)COOH	NH_2OR'
RCH(SSH)COOH	CS_2
RCH(COR')COOH	R'COCl
RCHC=O	CH_2CH_2
$\quad\Big|\quad\underset{}{\diagdown}O$	$\underset{O}{\diagdown\diagup}$
CH_2CH_2	

[a]Reaction accompanied by decarboxylation.

Polyoxyethylene esters.   Carboxylic acids react with ethylene oxide and an alkaline catalyst or with polyethylene glycol and an acidic catalyst to give a range of ethylene oxide esters. The product may also contain diol and diester resulting from ester interchange. In this way the carboxylic acids (anionic surfactants) are converted to non-ionic surfactants.

$$RCOOH + CH_2\text{-}CH_2 \xrightarrow{\quad O \quad} RCO(OCH_2CH_2)_n OH \longrightarrow H(OCH_2CH_2)_n OH + RCO(OCH_2CH_2)_n OCOR$$

$$RCOOH + HO(CH_2CH_2O)_n H \longrightarrow RCOO(CH_2CH_2O)_n COR$$

Bibliography

Baumann, H. and H. Biermann (1994) Oleochemical surfactants today, *Elais*, **6**, 49–64.

Christie, W.W. (1993) Preparation of ester derivatives of fatty acids for chromatographic analysis, in *Advances in Lipid Methodology – Two* (ed. W.W. Christie), The Oily Press, Dundee.

Dieckelmann, G. and H.J. Heinz (1988) *The Basis of Industrial Oleochemistry*, Peter Pomp, Essen.

Graille, J. (1993) Possible applications of acyltransferases in oleotechnology, *Lipid Technol.*, **5**, 11–16.

Gunstone, F.D. *et al.* (eds) (1994) *The Lipid Handbook*, 2nd edn, Chapter 10, Chapman and Hall, London.

Henkel (1982) *Fatty Alcohols, Raw Materials, Methods, Uses*, Henkel, Dusseldorf.

Johnson, R.W. and E. Fritz, (1988) *Fatty Acids in Industry*, Chapters 8 and 10, Dekker, New York.

Leyson, R. (1994) Renewable fuels in Europe, *Lipid Technol.*, **6**, 54–56.

Malcata, F.X. *et al.* (1990) Immobilised lipase reaction for modification of fats and oils – A review, *J. Amer. Oil Chem. Soc.*, **67**, 890–910.

Murayama, T. (1994) Evaluating vegetable oils as a diesel fuel, *INFORM*, **5**, 1138–1145.

Owusu-Ansah, Y.J. (1994) Enzymes in lipid technology and cocoa butter substitutes, in *Technological Advances in Improved and Alternative Sources of Lipids* (eds B.S. Kamel and Y. Yakuda) Blackie, London.

Pryde, E.H. (ed.) (1979) *Fatty Acids*, Chapters 11 and 12, American Oil Chemists' Society, Champaign, USA.

Schuchardt, U. and O.C. Lopes (1988) Tetramethylguanidine catalyzed transesterification of fats and oils: A new method for rapid determination of their composition, *J. Amer. Oil Chem. Soc.*, **65**, 1940–1941.

Staat, F. (1993) Vegetable oil methyl esters as a diesel substitute, *Lipid Technol.*, **5**, 88–61.

9 Dietary fats and nutrition

9.1 Storage and structural fats

Dietary fats of animal or vegetable origin are classified as 'visible' (adipose tissue, milk fat, seed oils) or 'invisible' (derived from animal or vegetable membranes). They are mainly triacylglycerols along with minor amounts of phospholipids, glycolipids, sterol esters, and vitamins. Fats are the richest source of energy on a weight basis and excess of fat beyond daily energy requirement is laid down as reserve depot fat, usually after some structural modification. Storage fat is to be found in adipose tissue, in milk fat, in the flesh of fatty fish, and in seed lipids. Fats in the adipose tissue are mainly triacylglycerols of saturated and unsaturated acids. The composition of this fat is not very critical and tends to reflect dietary intake (exogenous fat). Fat made *in vivo* from protein or carbohydrate is described as endogenous fat. The lipid composition of milk fat is more critical and is generally rich in essential fatty acids required by the neonate. It is, however, influenced by maternal diet. When storage fat is mobilized and metabolized it is oxidized to carbon dioxide and water with liberation of energy (section 7.4.2).

Lipids also play a significant structural role when they serve mainly as barriers such as skin and biological membranes. In the latter they are mainly phospholipids, glycolipids, or cholesterol with hydrophobic and hydrophilic segments. The fatty acid composition of these materials is more critical and they do not differ greatly from one species to another. Also their composition is less dependent on the nature of the dietary intake.

Dietary fat is the necessary source of essential fatty acids of the n-6 and n-3 types. These are needed for the maintenance of biomembrane integrity and for metabolism to the various products of the eicosanoid cascade (section 7.4.5). Some dietary fats are also required sources of the fat-soluble vitamins, i.e. A, D, E, and K.

Finally, fats affect the texture of food, make it more palatable, and may contribute to food flavour in both desirable and undesirable ways.

In the total picture of food intake it should be noted that (a) blood glucose must be maintained within very narrow limits since the brain cannot use free fatty acid as food, only glucose; (b) excess of carbohydrate is stored first as glycogen mainly in the liver and muscle and then as fat in adipose tissue; (c) protein supply must be at an adequate level, but excess of protein is converted to fat; and (d) when energy is required it is obtained first by mobilization of glycogen reserve, then from fat, and finally from protein.

9.2 Digestion, absorption, and transport

The digestion of fat starts in the mouth, continues in the stomach, and is completed in the small intestine. One of the problems involved in fat digestion, absorption, and transport, is that this material is insoluble in aqueous solutions. The products of digestion are more hydrophilic than ingested fat so they can be more easily dispersed. Lipids are also incorporated into lipoprotein complexes for transport through aqueous solution.

In the stomach fat is liberated from the food and churned into a coarse emulsion. In the small intestine this emulsion becomes associated with bile salts (amphiphilic compounds), colipase, and lipase (mainly from the pancreas). The system also has a requirement for calcium. Lipolysis liberates fatty acids from the sn-1 and sn-3 positions of triacylglycerols, and also from phospholipids, and cholesterol esters. Micelles are formed containing 2-monoacylglycerols, lysophospholipids, cholesterol, and free acids. About 90% of the triacylglycerols are absorbed in this way, but only about one half of the cholesterol esters. Short and medium-chain acids are not incorporated into these micelles but are rapidly absorbed and find their way via the portal vein into the blood vessels supplying the liver where they are quickly metabolized. It is for this reason that medium-chain triacyl-glycerides (MCT) are used therapeutically for patients unable to absorb long-chain acids.

The lipid hydrolysis products (monoacylglycerols, lysophospholipids, cholesterol, and fatty acids) pass into the enterocytes where lipids are resynthesized and assembled into the chylomicrons. These pass into general circulation via the thoracic duct and are transported in the bloodstream to the liver (roughly 30%), fat deposits (around 30%), and the musculature and other organs (about 40%). The bloodstream also contains triacylglycerols and free acids (as albumin complexes) coming from the liver and the fat depots. During fasting, free acids from the fat depots, circulating as albumin complexes, are removed by peripheral tissues where they are oxidized or by the liver where they are oxidized or converted to triacylglycerols before being returned to the blood.

Fats are transported through the bloodstream to and from sites of metabolism and storage as chylomicrons or other lipoprotein particles as shown in Table 9.1 these differ in the proportions of their major components (triacylglycerols, phospholipids, cholesterol, and proteins). Dietary fat is transferred as free acid to adipose tissue where it is reconverted to triacylglycerols. Endogenous fat, made mainly in the liver

Table 9.1 Human plasma lipoproteins

	Chylomicrons	VLDL[a]	LDL[a]	HDL[a]
protein[b]	2	7	20	50
triacylglycerol[b]	83	50	10	8
phospholipid[b]	7	20	22	22
cholesterol[b,c]	8	22	48	20
density range	<0.95	0.95–1.006	1.019–1.063	1.063–1.210

[a]VLDL (very low density lipoproteins), LDL (low density lipoproteins), and HDL (high density lipoproteins).
[b]% particle mass.
[c]Free and esterified.
Adapted with permission from *The Report of the BNF Task Force*, p. 29 (for full details see Bibliography at the end of this chapter).

but also in other organs, is exported as VLDL (very low density lipoproteins) into plasma. Cholesterol is carried to peripheral tissue in LDL (low density lipoprotein) and is returned to the liver in HDL (high density lipoprotein) which acts as scavenger for cholesterol.

Sucrose polyesters (section 8.3.4) are esters of sucrose with 6–8 acyl chains per sucrose molecule. These compounds are not hydrolysed by pancreatic lipase and are therefore not absorbed. They are excreted unchanged and thus have a very low or zero caloric value. They have been proposed as fat replacers in certain foods but do not yet have government approval in Europe or North America. Since they act as solvents for cholesterol and fat-soluble vitamins it may be necessary to supplement the latter when sucrose polyesters are ingested.

9.3 The role of fats in health and disease

For each individual there is an intake of fat (or other nutrient) which is appropriate to keep that individual in good health. Deviation from this optimum level in terms of having either too much or too little leads to some deterioration of health if not corrected over a suitable time scale. Small deviations in either direction, although slow to produce an observed effect, may nevertheless lead eventually to some undesired change. Large shifts from the optimum may result in serious ill health and, ultimately, in death.

Questions have been raised at various times about the safe levels of long-chain monoene acids such as erucic, of *trans* unsaturated acids (present in dairy fats, in partially hydrogenated oils, and in some processed oils), of saturated acids, and even of total fat. There is wide agreements that dietary fat in terms of both quality and quantity can influence early development (i.e. during pregnancy and in the neonate), coronary heart disease, plasma lipids, atherosclerosis, thrombogenesis, blood pressure, cardiac arrhythmias, cancer, diabetes, skin diseases, and disorders of the immune system. As a consequence of many investigations dietary recommendations have been made (a) for patients with these conditions and (b) for the general population. The latter are not meant to apply to pregnant and lactating females nor to children under five all of whom have their own special requirements. Such recommendations made by many committees in different countries are based on animal experiments, on epidemiological studies, and on a few studies with human volunteers. While there is a broad similarity between the various natural recommendations nevertheless the conclusions have been questioned by some individuals. It is clear, of course, that these disease conditions have many causes and even when fat plays a significant role it is by no means the only factor to be considered when choosing a healthy diet and life style.

Despite the recommendations that have been made few individuals have a quantitative idea of their dietary intake and it is interesting to note the difference between the mean intake (for the UK) and the figures recorded at the two extremes of the range. These lie between around one half and twice the mean value. The figures set out in Tables 9.2a and 9.2b are to be compared with recommended levels of fat intake.

There have been many sets of dietary guidelines in recent years. In general these indicate a reduction in total intake from 40–45% of total energy to around 30–35% and this is to be achieved mainly through a reduction of saturated acid from 20% to 15% or less. Some authorities suggest about 10% each (energy) of saturated, monounsaturated, and polyunsaturated acids. Present levels are around 16, 12 (and 2% *trans*) and 6%, respectively (UK adults). Recent studies indicate that this classification of acids is over-simplified. Among polyunsaturated acids there must be an appropriate balance between *n*-6 and *n*-3 members and perhaps also between C_{18} and $C_{20/22}$ polyene acids. It is also true that all saturated acids do not behave identically: some are less desirable while others are almost

Table 9.2a Daily intake of fat (grams) in the UK

Subject	Fat	sfa[a]	mufa[a]	pufa[a] *n*-6	pufa[a] *n*-3
men					
mean	102[b]	42	31	13	0.7
range[c]	50–156	19–69	16–50	5–30	0.4–1.3
women					
mean	74[b]	31	22	10	0.7
range[c]	31–124	12–55	10–37	3–21	0.4–1.4

[a]sfa etc. saturated fatty acids, monounsaturated fatty acids, polyunsaturated fatty acids.
[b]Equivalent to 40% of food energy.
[c]Values for the lower and upper 2.5%.
Adapted with permission from M.I. Gurr, *Role of Fats in Food and Nutrition*, 2nd edn, Chapman and Hall, London, 1992, p. 98.

neutral in their effects. Recent recommendations for the dietary intake of unsaturated acid for healthy adults are summarized in Tables 9.3 (*Report of the BNF Task Force*) and 9.4 (*Report of the Cardiovascular Review Group Committee on Medical Aspects of Food Policy, (UK)*).

Table 9.2b Daily intake of fat (grams and % total energy) for British adults (16–64 years)

Subject	Fat	sfa	mufa	*trans*	pufa n-6	pufa n-3
men						
mean (g)	102	42	31	5.6	13.8	1.9
energy (%)	37.6	15.4	11.6	2.0	5.1	0.7
women						
mean (g)	74	31	22	4.0	9.6	1.3
energy (%)	39.2	16.5	11.8	2.1	5.1	0.7

Abbreviations as for Table 9.2a.
J. Gregory, K. Foster, H. Taylor, and M. Wiseman (1990), *Dietary and Nutritional Survey of British Adults*, HMSO, London.

Table 9.3 Recommended dietary intake of unsaturated acids for healthy adults

Acid	% total energy avg population	Safe ranges		
		% total energy all	g/day men[b]	g/day women[c]
linoleic	6	3–10	8–26	6–20
α-linolenic	1	0.5–2.5	1–6	1–5
EPA + DHA[a]	0.5	0–2.0	0–5	0–4
all PUFA[a]	7.5	3.5–14.5	9–38	7–29
MUFA[a]	12	0–20	0–51	0–39

[a]EPA = eicosapentaenoic acid, DHA = docosahexaenoic acid, PUFA = polyunsaturated fatty acids, MUFA = monounsaturated fatty acids.
[b]Based on a 75 kg man consuming 2550 kcal/day.
[c]Based on a 60 kg man consuming 1940 kcal/day.
Adapted with permission from *The Report of the BNF Task Force*, p. 157 (for full details see Bibliography at the end of this chapter).

Table 9.4 Recommended dietary intake of fatty acid type for healthy adults

saturated acids	reduce to not more than 10% (energy)
cis-monoene acids	no specific recommendations but reductions in other types of acids could (if necessary) be replaced by acids of this type
trans-acids	no increase over the present level of about 2% (energy)
n-6 polyene acids	no further increase over the present level of 6% (energy)
n-3 polyene acids	double the present intake from 0.1 to 0.2 g/day
total fat	about 35% with more *cis*-monoene acids at the expense of saturated acids

Nutritional Aspects of Cardiovascular Disease, (1994) Report of the Cardiovascular Research Review Group Committee on Medical Aspects of Food Policy, HMSO, London.

Bibliography

Ashwell, M. (ed.) (1993) *Diet and Heart Disease – A Round Table of Factors*, British Nutrition Foundation, London.

Diplock, A.T. *et al.*, (eds) (1993) *Antioxidants, Free Radicals and Polyunsaturated Fatty Acids in Biology and Medicine*, IFSC, Lystrup, Denmark.

Emken, E.A. and H.J. Dutton (eds) (1979) *Geometrical and Positional Fatty Acid Isomers*, American Oil Chemists' Society, Champaign, USA.

Gregory, J. *et al.* (1990) *Dietary and Nutritional Survey of British Adults*, HMSO, London.

Gunstone, F.D. *et al.* (1994) *The Lipid Handbook*, 2nd end, Chapman and Hall, London.

Gurr, M.I. (1992) *Role of Fats in Food and Nutrition*, 2nd edn, Elsevier Applied Sciences, London.

Gurr, M.I. and J.L. Harwood (1991) *Lipid Biochemistry – An Introdution*, 4th edn, Chapman and Hall, London.

Jensen, R.G. *et al.* (1992) Lipids in human milk and infant formulas, *Ann. Rev. Nutrition*, **12**, 417–441.

Kamel, B.S. and Y. Yakuda (eds) (1994) *Technological Advances in Improved and Alternate Sources of Lipids*, Blackie, Glasgow.

Korsanger, H. (1986) *Dietary Trans Unsaturated Fatty Acids and Cardiovascular Disease*, A review submitted to DHSS by the Association of Fish Meal Manufacturers, London.

Lands, W.E.M. (1986) *Fish and Human Health*, Academic Press, London.

Larsson, K. (1994) *Lipids – Molecular Organisation, Physical Functions and Technical Applications*, The Oily Press, Dundee.

M.I. Gurr (1992) *Role of Fats in Food and Nutrition*, 2nd edn, Elsevier Applied Sciences, London.

Macdonald, I. (1987) *Trans Fatty Acids*, Report of the BNF Task Force, London.

Padley, F.B. *et al.* (eds) (1985) *The Role of Fats in Human Nutrition*, Ellis Harwood, Chichester.

Report of the British Nutrition Foundation's Task Force (1992) *Unsaturated Fatty Acids – Nutritional and Physiological Significance* Chapman and Hall, London.

Report of the Cardiovascular Review Group Committee on Medical Aspects of Food Policy (1994), *Nutritional Aspects of Cardiovascular Disease*, HMSO, London.

Robbelen, G. *et al.* (eds) (1989) *Oil Crops of the World – Their Breeding and Utilisation*, McGraw-Hill, New York.

Simopoulos, A.P. *et al.* (1986) *Health Effects of Polyunsaturated Fatty Acids in Seafoods*, Academic Press, London.

Vergroesen, A.J. and M. Crawford (eds) (1989) *The Role of Fats in Human Nutrition*, 2nd edn, Academic Press, London.

Yamanaka, W.M. *et al.* (1981) Essential fatty acids deficiency in humans, *Prog. Lipid Res.*, **19**, 187–215.

10 Practical applications of oils and fats

10.1 Introduction

As indicated in section 3.1 the 80–90 million tonnes of oil and fat produced annually are used mainly (80%) as human food. Dairy fat is consumed as milk or converted to butter, cheese, and other foods. Vegetable oils and animal fats are used, sometimes after appropriate modification, as fat spreads, cooking fats, frying oils, salad oils, mayonnaise, ice-cream, etc. A further 6% is used as animal feed and the remainder (14%) is converted to a range of oleochemicals of which over 90% are used as surfactants of one kind or another. The wide distribution of uses of such compounds is indicated later in this chapter. Some numerical data are given in Table 10.1.

10.2 Butter

Butter is a water-in-oil emulsion consisting of 80–82% fat and 18–20% of an aqueous phase which usually contains added salt. It is made from cow milk (3–4% fat) which is converted first to cream (30–45% fat) by centrifuging and then to butter (80–82% fat) by churning and kneading. During churning there is a phase-inversion from an oil-in-water to a water-in-oil emulsion. Annual world production is around 7 million tonnes.

The composition of milk fat changes in spring and autumn depending on the dietary intake. The fat is mainly triacylglycerols (97–98%) along with free acids and some monoacylglycerols and diacylglycerols. Also present are cholesterol (0.2–0.4%), phospholipids (0.2–1.0%, mainly phosphatidylcholine, phosphatidylethanolamine, and sphingomyelin), and traces of carotenoids, squalene, and vitamins A and D.

Milk fat is characterized by the presence of short and medium chain acids and of *trans*-monoene acids. It is one of the most complex and widely studied natural lipids and has been reported to contain over 500 acids of which about 15 are major components. Some numerical data (on a % wt basis) are presented in Table 10.2. Most butters contain 63–70% of saturated acids, 28–31% of monoene acids (including 4–8% with *trans* unsaturation) and 1–3% of polyene acids.

Trans-acids, produced by biohydrogenation of dietary lipids, are significant components of all ruminant milk fats (about 4–8%). They are

Table 10.1 Annual production (millions of tonnes) of oleochemicals (1993)

soap	7.1		fatty acids	2.4
oleochemicals	4.5	——	methyl esters	0.2
lubricants and coatings	0.5		alcohols	0.5
	——		amines	0.4
	12.1		glycerol	0.6
	——		other	0.4

Table 10.2 Major fatty acids (% wt) in cow milk fat

Fatty acid	June[a]	December[a]	Average[b]
4:0	4.2	3.5	3.6
6:0	2.5	2.2	2.2
8:0	2.3	1.1	1.2
10:0	2.2	2.6	2.8
12:0	2.4	2.8	2.8
14:0	9.0	10.6	10.1
16:0	22.1	26.0	25.0
18:0	14.3	11.6	12.1
20:0	2.6	2.3	2.1
18:1	30.4	24.8	27.1
18:2	1.2	2.8	2.4
Other	6.8	9.7	8.6

[a]Average of 4–8 samples.
[b]Average of 108 samples over 12 months.
Adapted with permission from *The Lipid Handbook*, 2nd edn (1994) p. 149.

mainly C_{16} and C_{18} monoene acids of which *trans*-vaccenic acid (11*t*-18:1) is the major component. A thorough study of French butter gave an average value for *trans*-acids of 3.8% and indicated that the average consumption of *trans*-acids from dairy sources in EU countries is 1.2 g per person per day. The range is 0.6–1.7 in different countries depending on the level of butter consumption.

Because of the wide distribution of chain length very different numbers are obtained when the results are expressed on a % mol basis. Butyric (butanoic) acid at about 4% (wt) corresponds to about 8.5% (mol). It is esterified predominantly at the *sn*-3 position and might be expected to be present in up to 25% of the triacylglycerols. The triacylglycerol composition is very complex but some information has been gained by gas chromatography and, more successfully, by HPLC. The latter technique generally gives 40–50 peaks each of which still consists of several components. Some of the major components are given in Tables 10.3 and 10.4. Compounds which contribute to the characteristic flavour of butter include γ- and δ-lactones (from 4- and 5-hydroxy acids), ketones, aldehydes, and short chain acids.

Table 10.3 Major triacylglycerols (% wt) in cow milk fat

Acyl groups[a]	Min[b]	Max[b]	Average[b]
30	0.8	1.6	1.1
32	1.6	3.8	2.6
34	3.9	7.9	6.0
36	8.2	13.8	10.8
38	10.0	14.0	12.5
40	8.6	10.8	9.9
42	5.1	8.7	6.9
44	4.2	8.9	6.5
46	5.2	9.7	7.3
48	7.9	10.7	9.1
50	8.2	13.4	11.3
52	5.2	15.5	10.0
54	1.5	11.1	5.0

[a]This number represents the total number of carbon atoms in three acyl chains and each group contains several glycerol esters. This distribution is bimodal. The C_{38} group contains one short chain and two longer chains (e.g. 4, 16, and 18), while the C_{54} group have mainly three C_{18} chains.
[b]Based on 755 samples from winter and summer feedings.
Adapted with permission from *The Lipid Handbook*, 2nd edn (1994) p. 152.

Table 10.4 Typical triacylglycerols (% wt) in cow milk fat

% wt		(%)
W[a]	S[a]	Glycerol esters[b]
7.8	6.0	BPO (65), CMP (30)
8.1	4.7	BPP (75), BMSt (15)
4.2	4.3	PPPo (10), MPO (75), LStO (10)
5.9	7.6	PPO (75), MStO (20)
2.6	4.5	PStO (95)

[a]These figures relate to the five largest fractions (>4%) from 49 separated by HPLC. W = Winter, S = Summer.
[b]Each glycerol ester includes all isomers containing the three acids indicated. B = butyric (4:0), C = caproic (6:0), M = myristic (14:0), P = palmitic (16:0), Po = palmitoleic (16:1), St = stearic (18:0), O = oleic (18:1).
K. Weber, E. Schulte, and H.-P. Thier (1988), *Fat Sci. Technol.*, **90**, 389–395 (this paper contains similar data for human milk fat).

Consumption of butter has declined in recent years as a consequence of its high price compared with alternative spreads, its poorer spreadability especially from the refrigerator, its high fat content with consequent high calorific value, and its relatively high level of saturated acids and of *trans*-monoene acids which are considered less desirable on nutritional grounds. On the other hand it has a characteristic and desirable flavour and mouth feel. Some of these problems have been overcome through legal changes and technological developments.

Butters with lower fat levels have been produced and present proposals for EU countries require that these be designated as butter (80–90% fat), three-quarter fat butter (60–62% fat), half fat butter (39–41%) and dairy spreads (other fat levels). These modified products will have reduced calorific value. Butter of modified composition is also produced from milk with changed fatty acid composition resulting from adjustment of the dietary regimen of the cows. Milk fat and butter fat enriched in monoene or polyene acids can be obtained in this way. It is now legal in many countries to blend butter and vegetable oils. This lowers the content of saturated acids which, in turn, influences melting behaviour and spreadability. Finally, fractionation of anhydrous milk fat gives a hard stock ('stearic acid fraction') and an oleic-rich fraction. These can be used on their own or blended to widen the use of butter fat (Table 10.5).

10.3 Margarine

Invented by Mège-Mouriès in 1869, margarine is now produced worldwide to the extent of 9 million tonnes per annum. This product was originally a cheap substitute for butter but it is now a product in its own right with some advantages over butter. As already indicated (section 10.2) attempts are being made to make butter more like margarine, although margarine manufacturers have been trying to introduce a butter-taste into their products. The main advantage of table margarine is that it can be spread at refrigerator temperatures. It can also be made with lower levels of saturated acids and higher levels of polyene acids compared with butter, and fat spreads can be produced with fat levels varying from 80% to 9–12%. It is expected that in EU countries the products will be designated according to their fat levels as margarine (80–90% fat), three-quarter fat margarine (60–62%), half fat margarine (39–41%), and fat spreads (other levels). Margarine often contains significant levels of *trans*-acids resulting from partial hydrogenation (section 7.1.1) though some products are available which contain little or no *trans*-acids.

Table 10.5 Uses of anhydrous milk fat (AMF) and its fractions

Butter fat components	Uses
AMF	cakes
AMF + oleic fraction	cookies and biscuits, butter cream
AMF + stearic fraction	fermented pastries, puff pastry
oleic fraction	ice cream cones, waffles, butter sponges, chocolate for ice cream bars

W. de Greyt and A. Huyghebaert (1993), *Lipid Technol.*, **5**, 138–140.

Margarine is made from appropriate oils and fats (soybean, rape, sunflower, cottonseed, palm, palm kernel, coconut) which may have been fractionated, blended, hydrogenated, and/or interesterified. Fish oil (hydrogenated or not) may also be included. Other ingredients include surface-active agents, proteins, salt, and water along with preservatives, flavours, and vitamins. The rapeseed oil used for this purpose must be low-erucic and there is a legal requirement that erucic acid be less than 5%. Usually it is much below this limit.

Margarines are made by cooling water-in-oil emulsions in scraped wall heat exchangers in which fat crystallization is initiated (nucleation) and the emulsion drop size reduced. This is followed by a maturing stage in working units during which crystallization approaches equilibrium. However, a considerable degree of fat crystallization is completed after the product has been packed. There are thus three basic steps: emulsification, crystallization of the solid part of the fat phase, and plastification of the crystallized emulsion.

Depending on the fatty acid composition and the consequent melting behaviour various products can be obtained such as soft (tub) and stick (packet) margarines. The latter is further divided into soft stick margarines used as spreads and hard stick material used for baking products. These have the approximate levels of saturated, monoene, and polyene acids shown in Table 10.6. A significant portion of the monoene acids have *trans* configuration.

Typical products will have a solid fat content (SFC) of 30–40% at 5°C to give good spreadability at refrigerator temperature, 10–20% at 10°C for good stand-up properties at room temperature, and less than 3% at 35°C for clean melting in the mouth.

As indicated in sections 6.1.3 and 6.1.4 triacylglycerols can crystallize in α, β′ and β forms. It is desirable that the crystals in margarine be in the β′ form and remain in this form. β′-crystals are small and can entrap a high level of liquid oil giving firm products with good texture and mouth feel. This is achieved by incorporating oils which crystallize in the β′ form such

Table 10.6 Approximate fatty acid composition of spreading fats

Fat	Saturated	Monoene[a]	Polyene
butter	63–70	28–31	1–3
stick (packet)	{ 18–21[b]	45–66[b]	14–35[b]
	{ 29–40[c]	46–52[c]	9–19[c]
soft (tub)	17–19	35–52	29–48

[a]US stick margarines contain 17–36% of *trans*-acids and soft margarines contain 10–18% of *trans*-acids.
[b]Vegetable oils only.
[c]Vegetable oils and animal fats.

as cottonseed, palm, tallow, modified lard, and hydrogenated fish oils. In some products sorbitan tristearate (0.3%) is added to inhibit conversion of β' to β crystal form. Oils crystallizing in the β' form generally have marked chain length diversity. With vegetable oils this implies appropriate levels of palmitic acid especially in the sn-1 and/or sn-3 positions, along with the predominant C_{18} acids. Palm oil and partially hydrogenated palm oil or palm olein are particularly efficient in this respect. In contrast to the β'-crystals, crystals in the β form are usually large and have a grainy mouth-feel. Structured lipids such as medium chain triglycerides, Caprenin, Salatrim, and Olestra have been discussed in section 2.9.7.

10.4   Other food uses

For frying, fats serve both as a heat medium and as a source of flavour, nutrition, and texture, since some of the fat is absorbed by the food being fried. In household use where the fat is not kept for a long time and is often discarded after limited use, requirements are not strict and liquid oils are widely used. In commercial establishments it is important to have good oxidative stability, high smoke point, minimum colour darkening, and good heat transfer. Ideally turnover should be as high as possible. For domestic use it is convenient for frying oils to be liquid at ambient temperature and even at refrigerator temperature so that they can be poured easily. For commercial/industrial frying the oil should be pourable/pumpable. Oils may be selected for their distinctive flavour (corn oil, olive oil, and tallow) or may be used in a refined bland form (cottonseed, groundnut, rapeseed, soybean, and palm olein), sometimes after partial hydrogenation.

Salad oils have similar requirements but it is more essential that the oil remain entirely liquid at ambient temperature.

10.5   Food-grade surfactants and emulsifiers

Surfactants and emulsifiers suitable for use in the food industry are produced to the extent of around 200 000 tonnes per annum. This demand is increasing with the growing use of convenience foods and with the demand for increased shelf-life.

These materials are amphiphilic lipids and are most often partial esters of medium and long-chain acids (C_{12}–C_{22}) with polyhydric alcohols such as glycerol, propylene glycol, sucrose, sorbitol/sorbitan, etc. Monoacyl-glycerols are by far the largest component (around 70%). Typically these compounds are used to stabilize water in oil emulsions such as margarine, to promote aeration of emulsions as in ice cream, and to extend the shelf-life of bread and other cereal products. Phospholipids are sometimes used for these purposes also.

Monoacylglycerols are most often made by glycerolysis of natural triacylglycerol mixtures in the presence of an alkaline catalyst (180–230°C, 1 h) as already discussed in section 8.3.3. Fat and glycerol (30% by wt) will give a mixture of monoacylglycerols (around 58%, mainly the 1-isomer), diacylglycerols (about 36%), and triacylglycerols (about 6%). This mixture can be used in this form or it can be subjected to high-vacuum thin-film molecular distillation to give a monoacylglycerol product (around 95%) with only low levels of diacylglycerols, triacylglycerols and free acids. The oils most commonly used include lard, tallow, soybean, cottonseed, sunflower, and palm and palm kernel in hydrogenated or non-hydrogenated form. Glycerol monostearate (GMS) is a commonly used product of this type.

Attempts are being made to effect this glycerolysis reaction with an enzymic catalyst (section 8.3.3).

The properties desired in a monoacylglycerol for some specific use may be improved by acylation of one of the free hydroxyl groups. This is achieved by reaction with acid (lactic, citric) or acid anhydride (acetic, succinic, diacetyltartaric) and the major products have the structures shown in Table 10.7. These are represented as stearates although other medium and long-chain acids can be used. As an example of the special properties of these compounds the so-called aceto-glycerides are stable in the α-crystal form and show unusual flexibility and film-forming properties. They can be stretched to eight times their original length.

Table 10.7 Derivatives of monoacylglycerols used as food grade emulsifiers

$$CH_2OCO(CH_2)_{16}CH_3$$
$$|$$
$$CHOH$$
$$|$$
$$CH_2OCOR$$

R	Name	Reagent
CH_3	acetate	acetic anhydride[a]
$CH(OH)CH_3$	lactate[b]	lactic acid
CH_2CH_2COOH	succinate	succinic anhydride[a]
$[CH(OAc)]_2COOH$	diacetyltartrate	diacetyltartaric anhydride
$\quad\quad\quad$ OH $\quad\quad\quad\ \|$ $HOOCCH_2CCH_2COOH$ $\quad\quad\quad\ \|$ $\quad\quad\quad$ COOH	citrate	citric acid

[a]React monoacylglycerol and anhydride at around 120°C for 1–2 h.
[b]The product will also contain molecules with 2–6 lactic acid units through oligomerization of this acid.

Propylene glycol ($CH_3CHOHCH_2OH$) also reacts with fatty acids to give mixtures of mono (about 55%, mainly 1-acyl) and diacyl esters (around 45%). A 90% monoacyl fraction can be obtained by molecular distillation.

Other compounds include the partial esters of polyglycerols (a polyether with 2–10 glycerol units but mainly 2–4 units), sorbitan (and its polyethylene oxide derivatives), the 6-monoacylate of sucrose, and stearoyllactate (usually as the sodium or calcium salt).

10.6 Non-food uses of fatty acids and their derivatives

The range of fatty acids and derivatives and their variety of non-food uses are very wide so that it is not possible to cover all these and this account presents only a selection of both major and minor uses. Most of these are based on the amphiphilic properties of these molecules, i.e. the combination within one molecule of hydrophobic (lipophilic) and hydrophilic (lipophobic) regions. This is illustrated in the simple case of a soap (Figure 10.1).

Surface-active agents, known as surfactants, fall into four groups depending on the nature of the polar head group:

anionic: e.g. sulphates, sulphonates (and carboxylates)
nonionic: e.g. ethoxylates of hydroxy compounds (phenols, alcohols, carboxylic acids) and amino compounds (amines, amides)
cationic: quaternary ammonium salts (quats), imidazoline salts
amphoteric: betaines, imidazolines

The structures of these compounds have been given in sections 8.6 and 8.7. Levels of consumption lie in the sequence anionic > nonionic > cationic > amphoteric. Consumption figures are often confusing since it is not always clear whether traditional soap has been included or not.

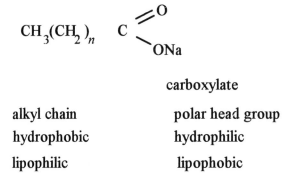

carboxylate

alkyl chain   polar head group
hydrophobic   hydrophilic
lipophilic   lipophobic

Figure 10.1 Hydrophilic and hydrophobic regions in a soap.

Surfactants generally contain one or more acyl groups which may be C_{12}, $C_{16,18}$ or C_{22}.

Surfactant properties include those listed in Table 10.8. Sometimes these operate in opposite directions and suitable materials or mixtures must be selected for a particular purpose. The major applications of surfactants are listed in Table 10.9. These compounds have been widely used in North America, western Europe, and Japan but demand is spreading through all developed and developing countries.

(i) *Nitrogen derivatives.* Amides, especially those derived from oleic acid and from erucic acid, are used mainly as antislip and antiblock additives for polyethylene film in which they are incorporated at a 0.1–0.5% level. This is a permitted food packaging material. They are also used as water repellents for textiles, mould-release agents, and in rubber goods and printing inks. 22 000 tonnes of high-erucic oil furnish about 7000 tonnes of erucamide. Ethanolamides are incorporated into dish-washing formulations.

Amines find use as corrosion inhibitors, mould-release agents, pigment-wetting agents, biocides, and for mineral flotation. They are also used to protect hygroscopic material, such as fertilizers, from atmospheric moisture. Quats are extensively used as fabric softeners, in the production of organomodified clays for drilling mud and thixotropic paints, as sanitizers in restaurants, dairies, and hospitals, and in water-based asphalt emulsions for road building and maintenance. Amine oxides are weak cationic

Table 10.8 Properties of surfactants

emulsification	defoaming	detergency
de-emulsification	water-repelling	sanitizing
wetting	dispersing	lubricity
foaming	solubilizing	emolliency

Table 10.9 Major applications of surfactants

Consumer products	Industrial products	
detergents	detergents and cleaners	food industry[b]
dishwashing agents	textiles and fibres	pulp and paper, inks
cleaning agents	mining and ore flotation	agrochemicals
personal products[a]	petroleum production	leather and skin
cosmetics	lubricants (for engines)	metal working
pharmaceuticals	paints, laquers, plastics	cement and concrete

[a]Soap, shampoos, bubble baths, toothpaste, shaving creams, shower gels, etc.
[b]Additives, cleaners, and biocides.

surfactants with detergent, foaming, antistatic, and antiseptic properties and find wide use in shampoos and bubble baths. Polyethyleneoxide amines are used as foaming and frothing agents, corrosion inhibitors, mud-drilling additives, and textile-finishing agents.

(ii) *Alcohols* (section 8.6).   These are made from appropriate oils and fats mainly in the C_{12} or C_{16}–C_{18} ranges or from ethylene (C_{12} range). The shorter-chain members (C_6–C_{10}) are used in ester form (e.g. phthalates) as plasticizers for PVC etc. The C_{12} and $C_{16,18}$ alcohols are included in a wide range of surfactant formulations mainly as sulphate ($ROSO_2OH$) or ethoxylate ($RO(CH_2CH_2O)_nH$). Oleyl alcohol finds use as an emollient and emulsion stabilizer in cosmetic and pharmaceutical preparations.

(iii) *Polyethoxylated fatty acids* ($RCOO(CH_2CH_2O)_nH$).   These are non-ionic surfactants. They are included in formulations for cleaning clothes, dishes, hard surfaces, and metals and in textile processing. In the paper industry they are used to improve wet strength and in the deinking of newsprint. In agriculture they serve as emulsifiers for pesticides.

(iv) *Dibasic acids*.   Important aliphatic dibasic acids are made from petrochemicals (C_6, C_{12}) or from oleochemicals (C_9, C_{10}, C_{13}). They are used as plasticizers for PVC (as C_4–C_8 dialkyl esters), as lubricants for engines (C_8, C_{10}, C_{13} dialkyl esters), and can be made into polyamides, polyesters, and polyurethanes which are used as adhesives, coatings, fibres and nylons.

(v) *Dimer acids*.   These (and trimer acids) are used mainly for the production of polyamide resins (section 7.11). Reactions with diamines give neutral polyamides with no free amine groups which are used in printing inks and in the production of hot-melt adhesives for shoes and book-binding etc. Dimer acids also react with tri- and tetra-amines to give products with free amino groups capable of further reaction. These are used as curing agents for epoxy resins and give products used as adhesives and surface coatings. Dimer acids are adsorbed on metal surfaces where they act as corrosion inhibitors. They are also used in pipes for this purpose.

(vi) *Cosmetic and pharmaceutical uses*.   Fatty acid derivatives incorporated into cosmetics and personal care products such as skin cleaners, bath preparations, lotions, creams, toothpaste, shaving preparations, shampoos and soaps must be non-irritant especially to the eyes, chemically stable to hydrolysis by acid or alkali, have very low oral and percutaneous toxicity, be free of microbial contamination, and where relevant, must produce physically stable preparations. Various kinds of surface-active compounds are used including amines and cationic, non-ionic, and amphoteric materials. They are employed for their emollient or lubricating properties and for their ability to stabilize emulsions. Alkyl esters such as isopropyl myristate and salts of fatty acids (zinc, magnesium, sodium, aluminium) also find wide use.

(vii) *Metal-working fluids.* Polyethylene glycol esters, $C_8–C_{11}$ amine salts, fatty acid soaps (sodium, potassium, amine), amides from diethanol-amine, and dimer acids are used as components of metal-working fluids.

Diamides of the type $RCONH(CH_2)_nNHCOR$ where $n = 8–12$ and RCO represents $C_1–C_{18}$ acyl groups show good antirust properties for water-based cutting fluids. They also show good lubricant and antimicrobial properties.

(viii) *Textiles.* Long-chain amines, acids, alcohols, and compounds de-rived from these are extensively employed as textile processing aids. They serve as detergents, lubricants, antistatic agents, and softeners and also assist in dyeing processes.

(ix) *Pulp and paper making.* Tall oil is a useful by-product of the paper industry which itself uses significant amounts of fatty materials of various kinds for sizing and also as defoamers, lubricants, and flotation agents (in recycling).

(x) *Lubricants.* Vegetable oils are being considered as replacements for petroleum-based lubricants. This is an inversion of history because such oils were used for these purposes before the advent of the hydrocarbon lubricants. Today's demand for lubricants is so large that products based on vegetable oils can make only a small contribution and lubricants are probably still more than 97% petroleum based.

Nevertheless there is a growing demand for the oil-based products particularly in situations where there is rapid loss of lubricant, e.g. with chain-saws where the chain lubricant is lost on to the forest floor and in outboard engines (particularly on inland lakes and seas) where lubricant is spilled into the water. Under these conditions the shorter lubricant life of the vegetable oil is not a serious disadvantage and this is more than balanced by ecological factors. Compared with hydrocarbon lubricants those derived from vegetable oils have good technical properties and give fewer medical problems and allergies. They are biodegradable, non-toxic and non-carcinogenic, and come from a renewable resource. On the other hand they are usually more costly and show some limitations, such as a limited viscosity range and lower oxidative and hydrolytic stability, compared with the petroleum oils. Some of these problems can be overcome with suitable additives some of which can themselves be made from vegetable oils. For example, a telomer of molecular weight about 6000 made from a mixture of rape and linseed oils (or from other unsaturated vegetable oils) by heating at 300–320°C for 12–15 hours in a nitrogen atmosphere can be used to improve the viscosity index. Lubricants normally consist of a petroleum base-fluid and a wide range of additives to improve performance. New synthetic lubricants (synlubes) include a range of esters based on products of the petrochemical and/or oleochemical industry and may also need appropriate additives. While generally superior to the petroleum-based lubricants they are more expensive and their superior performance must be balanced against additional cost. Ester synlubes are of two kinds: diesters of dibasic acids such as adipic, azelaic, sebacic, or aromatic acids with C_8–C_{13} alcohols which are frequently branched-chain or esters of trihydroxymethylpropane or pentaerythritol with C_5–C_{10} straight- and branched-chain acids.

$ROCO(CH_2)_nCOOR$ $CH_3CH_2C(CH_2OH)_3$ $C(CH_2OH)_4$

diester   trihydroxymethylpropane   pentaerytritol

Synlubes are widely used in systems subject to large extremes of temperature such as aircraft engines and those running for long periods without interruption (compressors). There are many instances where the more expensive synlubes are more cost effective than the cheaper petroleum-based lubricants.

(xi) *Oilfield chemicals.* Fat-based materials are extensively used as oilfield chemicals. Sulphurized oils, the calcium and magnesium salts of fatty acids, dimer acids, and lecithin are important components of drilling mud, but the largest use is of amine salts of tall oil fatty acids and of dimer acids and of imidazolines based on diamines and triamines as corrosion inhibitors. Biocides which inhibit microbial activity include the hydrochloride of the amine shown:

$RCONH(CH_2CH_2NH)_2CH(Me)COOH$

(xii) *Inks.* Vegetable oil-based inks incorporate materials from soya, linseed, rape, or palm. They have several benefits over the more traditional petroleum-based products. These include, superiority in rub resistance, stability in the press, production of a deeper black, quicker adjustment to change of colour, and easier removal when recycling the paper. This is a niche market which could lead to a small but significant demand for fat-based products.

(xiii) *Nylons.* Undecenoic acid (from castor oil) can be converted to 11-aminoundecenoic acid which is the monomer required to produce the polyamide Rilsan with specific uses.

(xiv) *Use as a solvent.* Methyl esters obtained from soybean or other oils are the basis of a new industrial solvent used to remove grease, asphalt, tar, hydrocarbons, etc. The solvent is safe to handle and fully biodegradable.

(xv) *Factice.* This is made by heating vegetable oils and sulphur (brown factice) or sulphur chloride (white factice) and is used in the rubber industry. Mixtures of soybean oil with rapeseed or meadowfoam oil give good products.

(xvi) *Biodiesel.* This material, most commonly described as biodiesel, is also known as diester (from diesel ester), RME (rapeseed methyl esters), palm diesel, or by several other names.

The methyl esters of vegetable oils or animal fats can be used in partial or complete replacement of diesel fuel without any modification of the engine. Glycerol esters (the original oil or fat) can also be used but only in a modified engine such as the Elsbett engine produced in Germany. The esters are made by methanolysis with either acidic or basic catalysts (section 8.3.3) and have been produced for use as biodiesel from rape, sunflower, palm, coconut, and from animal fats (tallows), and even from spent frying oil.

The motivation for the production of biodiesel is partly agricultural and partly environmental. Their potential is being examined seriously in Europe, Malaysia, and elsewhere. In France, there is an aim to produce 500 000 tonnes of diester per year. This represents less than 5% of the country's consumption of diesel but will take the output from about 400 000 hectares of rapeseed. In another study it was concluded that 1 bushel of soybean could give 1.5 gallons of methyl ester at a cost of 1.5–2.0 US dollars per gallon. The ecological claims are that the material is not toxic, that it is completely biodegradable, and that, compared with diesel, it produces less sulphur compounds, less carbon monoxide, less polycyclic aromatic hydrocarbons, and less black smoke when used as a fuel. It has been shown that it produces more energy than is required for its inputs and that it does not add to total carbon dioxide since that produced on combustion is balanced by that used in the biological production of the seed oil. It is used mainly in admixture with diesel fuel by captive fleets such as taxis and buses especially in inner urban areas where the atmospheric conditions are poorest. It is difficult to calculate the true economic cost of all the factors involved but the production and use of biodiesel is unlikely to be economically viable without a subsidy or a reduction of fuel tax or carbon tax (if introduced) for conventional fuels. The amount of oil or fat available for the production of biodiesel is limited and can only have a small impact on the total demand for diesel fuel.

(xvii) *Use by the agricultural industry.* Vegetable oils and their derivatives are applied for a number of agricultural purposes. They can be used to fight crop pests, both in the growing crop and in the stored product. For example, pesticides are effectively delivered in the form of an emulsion made from rapeseed or other vegetable oil, an emulsifier, and the pesticide. They are also used to suppress dust in granaries and in flour mills.

Large hay bales kept in the open air can lose 25–40% of their nutritional value through spoilage. This can be markedly reduced if the hay bales are sprayed with rendered animal fat – either hydrogenated or not. The fat layer also increases the energy value of the hay for livestock.

The calcium salts of fatty acids (usually from palm fatty acid distillate – a by-product from the refining of palm oil) are used as components of animal feed, especially for lactating cows and sheep (ruminants). The salt passes unchanged through the slightly acidic rumen but is converted to free acid in the highly acidic abomasum. The free acids are thus released for metabolism as an efficient source of energy.

Bibliography

Ahmed, A. *et al.* (1994) How much energy does it take to make a gallon of soydiesel? *Report of the Institute for Local Self-Reliance – Environmentally Sound Economic Development*, Washington DC.

Applewhite, T.M. (1991) *World Conference on Oleochemicals into the 21st Century*, American Oil Chemists' Society, Champaign, USA.

Baumann, H. and M. Biermann (1994) Oleochemical surfactants today, *Elaies*, **6**, 49–64.

Berger, K.G. (1989) Functionality and interchangeability of fats, *Lipid Technol.*, **1**, 40–43.

Berger, K.G. (1993) Food product formulation to minimize the content of hydrogenated fats, *Lipid Technol.*, **5**, 37–40.

Charteris, W. and K. Keogh (1991) Fats and oils in table spreads, *Lipid Technol.*, **3**, 16–22.

Charteris, W. (1991) Monobutter: a fridge-spreadable butter, *Lipid Technol.*, **3**, 90–91.

De Greyt, W. and A. Huyghebaert (1993) Food and non-food applications of milk fat, *Lipid Technol.*, **5**, 138–140.

de Man, L. and J. de Man (1994) Functionality of palm oil in margarines and shortenings, *Lipid Technol.*, **6**, 5–10.

Dieckelmann, G. and H.J. Heinz (1989) *The Basis of Industrial Oleochemistry*, Peter Pomp, Essen.

Erhan, S.Z. and M.O. Bagby (1994) Polymerization of vegetable oils and their uses in printing inks, *J. Amer. Oil Chem. Soc.*, **71**, 1223–1226.

Erickson, D.R. *et al.* (eds) (1980) *Handbook of Soy Oil Processing and Utilization*, American Soybean Association and American Oil Chemists' Society.

Eyres, L. (1989) Milkfat product development, *Lipid Technol.*, **1**, 12–16.

Falbe, J. (ed.) (1986) *Surfactants in Consumer Products – Theory, Technology and Application*, Springer-Verlag, Berlin.

Flack, E. and N. Krog (1990) Emulsifiers in modern food production, *Lipid Technol.*, **2**, 11–13.

Garti, N. and K. Sato (eds) (1988) *Crystallisation and Polymorphism of Fats and Fatty Acids*, Dekker, New York.

Gunstone, F.D. *et al.* (eds) (1994) *The Lipid Handbook*, 2nd edn, Chapman and Hall, London.

Henkel, K. (ed.) (1982) *Fatty Alcohols, Raw Materials, Methods, Uses*. Henkel, Dusseldorf.

Hoffmann, G. (1989) *The Chemistry and Technology of Edible Oils and Fats and Their High Fat Products*, Academic Press, London.

Johnson, R.W. and E. Fritz (1988) *Fatty Acids in Industry – Processes, Properties, Derivatives, Applications*, Dekker, New York.

Karsa, D.R. (ed.) (1987) *Industrial Applications of Surfactants*, RSC, London.

Karsa, D.R. (ed.) (1992) *Industrial Applications of Surfactants III*, RSC, London.

Kaylegian, K.E. and R.C. Lindsay (1994) *Handbook of Milkfat Fractionation Technology and Application*, AOCS Press, Champaign, USA.

Larsson, K. (1994) *Lipids – Molecular Organisation, Physical Functions and Technical Applications*, The Oily Press, Dundee.

Leonard, C. (1994) Sources and commercial application of high-erucic vegetable oils, *Lipid Technol.*, **6**, 79–83.

Leysen, R. (1992) Non-edible applications of soybean oil, *Lipid Technol.*, **4**, 65–69.

Mang, T. (1994) Biodegradable lubricants and their future markets, *Lipid Technol.*, **6**, 139–143.

Moran, D.P.J. and K.K. Rajah (eds) (1993) *Fats in Food Products*, Chapman and Hall, London.

Murphy, D.J. (ed.) (1993) *Designer Oil Crops – Breeding, Processing, and Biotechnology*, VCH, Weinheim.

Nichols, B. (1993) The current market and legal status of butters, margarines and spreads, *Lipid Technol.*, **5**, 57–60.

Ohlson, R. (1993) Vegetable oils for non-food products, *Lipid Technol.*, **5**, 34–37.

Rajah, K. (1992) Recent technology for the manufacture of butter, margarine, and spreads, *Lipid Technol.*, **4**, 129–137.

Robbelen, G. *et al.* (eds) (1989) *Oil Crops of the World – Their Breeding and Utilisation*, McGraw-Hill, New York.

Rossell, J.B. and J.L.R. Pritchard (eds) (1991) *Analysis of Oilseeds, Fats, and Fatty Foods*, Elsevier, London.

Rubingh, D.N. and P.M. Holland (eds) (1991) *Cationic Surfactants – Physical Chemistry*, Dekker, New York.

Turner, G.P.A. (1988) *Introduction to Paint Chemistry and Principles of Paint Technology*, Chapman and Hall, London.

Vietmeyer, N.D. (1985) *Jojoba – New Crop for Arid Lands – New Raw Material for Industry*, National Academy Press, Washington.

Young, Y. (1990) The usage of fish oils in food, *Lipid Technol.*, **2**, 7–10.

Index

α-anions   221
acetate   23, 106
acetate–malonate pathway   24
acetylenic acids   10, 11, 159
acetylenic intermediates   30
acid anhydrides   214
acid chlorides   214
acidity   102
acidolysis   208, 209, 213
ACP, *see* acyl carrier protein
activated carbon   90
acyl carrier protein   25
acylation   47
acylglycerols   36, 52
acylpyrrolidide   154
alchornoic acid   15
alcohols   217, 238
alcoholysis   208, 209
alepramic acid   14
aleprestic acid   14
alepric acid   14
aleprolic acid   14
aleprylic acid   14
Aleurites fordii   10
alkanoic acids, crystal structure   130
alkylbenzoxazole   154
alkyl dimethyloxazole   154
alkyl ketene dimers (AKD)   215
alkyl polyglucoside   217
allenic acids   10, 11, 39
allenic epoxides   187
allylic bromination   194
alklylic hydroperoxides   162, 163, 165
almond oil   66, 67, 69
alternation   130
amides   218, 238
amine oxides   219, 220, 237
amines   219, 220, 237
amphiphilic lipids   234
amphoterics   219, 237
analytical procedures   100
   see also specific types of
anchovy oil   72, 74
anhydrous milk fat   96
animal fats   72
anionic surfactants   205, 237
anisidine value   103, 104
annual crops   61
*ante*iso acids   12, 26

anti-block agent   218, 237
antioxidants   84, 173, 175, 176
anti-slip agent   218, 237
Apis mellifera   85
apricot oil   66, 67
arachidic acid   5, 6
arachidonic acid   8, 9, 29, 184
   synthesis   31, 32
Arachis hypogaea   67
argentation tlc   19
arsenites   18
asclepic acid   7
Aspergillus niger   88, 97, 213
Aspergillus oryzae   213
auricolic acid   13, 14
autoxidation   162–167
avenasterol   84
avocado oil   67
aziridines   202

babassu oil   88
beef tallow   72
beeswax   84
behenic acid   5, 6
bentonitite   90
benzyl-*sn*-glycerol   54
benzylidene glycerol   48
Betapol   55, 214
BHA, *see* butylated hydroxyanisole
BHT, *see* butylated hydroxytoluene
biodiesel   241
biohydrogenation   161
biosynthesis   24, 44
blackcurrant oil   64, 65
bleaching   89, 90
Bligh and Dyer method   101
BOB, *see* oleodibehenin
boiling point   5
bolekic acid   11
borage oil   64, 65
Borago officinalis   65
borates   18
Borneo tallow   66
bovine milk phosphatidylcholines   122
branched-chain acids   12, 85
Brassica alba   79
Brassica campestris   68
Brassica napus   68
brassicasterol   84

bromination 193, 194
bromo acids 193
butanoic acid 4, 5, 25, 230
butter 63, 229, 233
butterfat 63, 71
butylated hydroxyanisole 176, 177
butylated hydroxytoluene 176
butyl peroxyester 215, 216
butyric acid 4, 5, 230
Butyrivibrio fibrosolvens 161
Butyrospermum parkii 66
by-products, oils as 61

caffeic acid 179
calendic acid 10
Calendula officinalis 10
calendula oil 80
campesterol 84
candelilla wax 84
Candida cylindricae 97, 213
Candida rugosa 207
canola oil 174
capelin oil 72, 74
caper spurge 79
Caprenin 56, 213
capric acid 5
caproic acid 5
caprylic acid 5
carbonyl compounds, infrared absorption
139
carboxyl group reactions 205, 220
carnauba wax 84
carnosic acid 179
carotenes 82, 179
carrot seed oil 146
Carthamus tinctorius 70
castor oil 13, 14, 62, 63, 65, 153
Catalpa ovata 10
catalpic acid 10
cationic surfactants 205, 237
Cephalocroton cordofanus 15
ceramides 36, 44, 46, 47
cerebroside, biosynthesis of 46
cerebrosides 36, 44
cetoleic acid 7, 8
chain elongation 26
chaulmoogric acid 14
chemical shifts
13C 144, 145, 146
1H 142
chicken fat 124
chlorination 193
chloro acids 193
chlorophyll 82
chocolate 65
cholesterol 83
oxidation 171
chromatographic refining 90

chromatography *see* specific forms of
chromatorods 106
chrysobalanic acid 10
Chrysobalanus icaco 10
chylomicrons 224, 225
citric acid 179
13C-nmr chemical shifts 144–146
CoA esters 25
cocoa butter 65, 66, 75, 120, 124, 137
polymorphism 138
substitute 213
coconut oil 62, 63, 65, 66, 174
glycerol esters 114
Cocus nucifera 66
cod liver oil 72, 74, 174
cod roe 83
columbinic acid 10
complex lipids 35
component acids 69, 106
compound formation 134
confectionary fats, crystal structure 137
conjugated polyene acids 9
Copernica cerifera 85
copra 66, 88
coriolic acid 15
corn oil 62, 63, 66, 69, 88, 174
Cornyebacterium acnes 94, 213
coronaric acid 15
cottonseed oil 62, 63, 66, 69, 76, 88, 174
cow milk fat 71
Crambe abyssinicum 79
Crambe oil 79
crepenyic acid 10, 11
Criegee zwitterion 190
crystal form of glycerol esters 134, 135,
233
crystallisation 17
crystallographic data 132
Cuphea sp. 78
cutins 43
cyclic acids 12
cyclisation 198
cyclopentene acids 13, 14, 26
cyclopropane acids 203
cyclopropene acids 181, 203

DAG, *see* diacylglycerols, synthesis of
dairy spreads 232
DCL (double chain length) structures
133, 136
de novo biosynthesis 23, 25
decanoic acid 5
degumming 89, 90
dehalogenation 195
dehydrated castor oil 14
dehydrocrepenyic acid 11
dehydrohalogenation 195
demospongic acids 10

densipolic acid 13, 14
deodorisation 89, 90
desaturation 27
deuterated acids, synthesis 31
DGDG, *see* digalactosyldiacylglycerols
DHA, *see* docosahexaenoic acid
diacyl peroxides 215, 216
diacylglycerols 37, 41
  crystal structure 136
  synthesis 50, 51, 55
diamides 218, 239
diamines 219, 220
diazomethane 208
dibasic acids 238, 240
diepoxystearic acid 186
dietary guidelines 226
dietary intake 226, 227
differential scanning calorimetry 129
digalactosyldiacylglycerols 36, 39
diglycerides, *see* diacylglycerols
dihomo-γ-linolenic acid 8, 29, 30
dihydroxystearic acid 189
dilatometry 129
dimer acids 173, 200, 238
dimer alcohols 217
dimerisation 200
dimorphecolic acid 15, 81
dimorphotheca oil 81
diols 189, 190
diphytanylglycerol 43
directed interesterification 97
distillation 17
docosahexaenoic acid 8, 9, 29, 208
docosanoic acid 5
docosapentaenoic acid 8, 28, 29
docosenoic acid 7
dodecanoic acid 5
double bond migration 92, 198
doxylstearic acid 138
DPA, *see* docosapentaenoic acid
dry fractionation 95

E numbers 177
ECL, *see* equivalent chain length
EDTA, *see* ethylene diamine tetra-acetic
  acid
EFA efficiency 28
egg lipids 82
egg phosphatidylcholines 122
eicosanoic acid 5
eicosanoid cascade 185
eicosanoid metabolites 29
eicosanoids 184
eicosapentaenoic acid 8, 9, 29, 203
eicosatetraenoic acid, synthesis 31
eicosenoic acid 7
Elaeis guinensis 68
elaidic acid 7

elaido glycerides 93
electron-impact mass spectroscopy 147, 148
electron spin resonance spectra 138
eleostearic acid 10
emulsifiers 234
endogenous fat 223
enzymes 97, 205
enzymic hydrolysis 41
enzymic interesterification 97, 212
enzymic oxidation, mechanism of 185
EPA, *see* eicosapentaenoic acid
epithiostearic acid 202
epoxidation 186, 187
epoxide 189
epoxidised oils 147
epoxy acids 15, 185–188
epoxy oils 81
epoxypropanol 48
epoxystearic acid 186
equivalent chain length 20, 21, 111
Erlanga tomentosa 15
erucic acid 7, 68, 79, 225
erythrogenic acid 11
essential fatty acids 223
ester waxes 43
ester–ester interchange 208
esterification 207, 208
estolide 201
ethanediol esters 43
ethanolamides 237
ether lipids 42
ether sulphate 217
ethoxylate 217
ethoxylated amides 218
ethylene diamine tetra-acetic acid 179
ethylene oxide adducts 219
Euphorbia antisphylitica 85
Euphorbia lagascae 15, 81
Euphorbia lathyris 79
even acids – polymorphism 132
evening primrose oil 64, 65
exocarpic acid 11
exogenous fat 223
exotic oils 66
extraction 87, 88

factice 241
FAS, *see* fatty acid synthetase
fat spreads 232
fats
  absorption 224
  daily intake 62, 226, 227
  digestion 224
  splitting 207
  transport 224
  use 241
fatty acid synthetase 25

biosynthesis 23
composition 73
fatty acids
  chemical synthesis 28
  cosmetic uses 238
  dietary intake 227
  identification 16
  isolation 16
  structure 3
fish oil 62, 63, 72, 73
Flacourtiaceae, seed fats 13
flavour 165
fluoro-oleic acid 16
fluoropalmitic acid 16
Folch method 101
food grade emulsifiers 235
Fourier transform infrared spectrometers 138
fractionation 95
Friedel–Crafts reaction 203
frying oils 234
Fuller's earth 90
furanoid acids 15

gadoleic acid 7
gallic esters 177
gangliosides 36, 44
gas chromatography 106, 111, 112, 113, 115, 116
Geotrichum candidum 97, 213
GLA, see γ-linolenic acid
glycerides 36
glycerol dibehenate oleate 55
glycerol dioleate palmitate 55
glycerol esters
  biosynthesis 46
  crystal structure 133
glycerol monostearate 235
glycerol triesters 133
glycerol tristearate 131
glycerolysis 208, 210, 234
glycerophosphoric acid 39
glycerophosphorylcholines 41
glycidol 48
Glycine max 70
glycosyldiacylglycerols 39
glycosylglycerides 36
gondoic acid 7
gorlic acid 14
Gossypium hirsutum 66
gourmet oils 66
GPC, see glycerophosophorylcholines
grapeseed oil 67
groundnut oil 62, 63, 67, 69, 76, 88
Guerbet acids 201, 217
Guerbet alcohols 201, 217

half-fat butter 232

half-fat margarine 232
half-hydrogenated intermediate 160
halogenation 192
hazelnut oil 67
HDL (high density lipoprotein) 225
helenynolic acid 10, 15
Helianthus annus 70
heptanal 14
herring oil 72, 73, 127, 174
hexadecanoic acid 5
hexadecenoic acid 7
hexahydroxystearic acid 189
hexanoic acid 5
high-performance liquid chromatography 106, 109, 110, 114, 117, 119
higher monoene acids 79
HighSun oil 71
^{1}H-nmr, chemical shifts 143
honesty seed oil 8, 79
hormelic acid 14
HPLC, see high-performance liquid chromatography
human milk fat 71, 73, 155, 214
hydnocarpic acid 14
hydrazine reduction 159, 161
hydroformylation 203
hydrogenated fats 147
hydrogenation 22, 89, 91, 157, 158, 160
hydrogenolysis 217
hydrolysis 206
hydroperoxide 166–168, 185, 191
hydrophilic regions of soap 236
hydrophilisation 96
hydrophobic regions of soap 236
hydroxy acids 13, 14, 81, 185
hydroxydecanoic acid 14
hydroxylation 189
hyproperoxide lyase 183

Iatroscan 106
illipe butter 66
imidazolines 219, 220
Impatiens balsamina 10
infrared spectroscopy 129, 138, 139
inks 240
interchangeability of oils and fats 76
interesterification 96, 97, 208, 210, 211, 213
invisible fat 223
iodolactonisation 193
isanic acid 11
isanolic acid 15
iso acids 12, 26
isopropyl esters 108
isopropylidine glycerol 48, 53, 55
isoricinoleic acid 14
isostearic acid 201
isotopically labelled acids, synthesis 32

Jacaranda mimosifolia 10
jacaric acid 10
Jessenia palm oil 79
jojoba oil 8, 43, 79, 84

kamlolenic acid 10
Koch reaction 203
kusum 6

laballenic acid 11
lactobacillic acid 12
lactones 85, 230
lamenallenic acid 11
Lanza process 95
lard 62, 63, 71, 72, 97, 174
lauric acid 5, 25
lauric oils 66, 73, 78
LDL (low density lipoprotein) 225
lecithin 90
Lesquerella fondleri 81
lesquerolic acid 13, 14, 81
leukotrienes 184
licanic acid 10
lignoceric acid 5, 6
Limnanthes alba 8, 80
Lindlar's catalyst 31
linola oil 80
linoleic acid, and ester 8, 9, 28, 29, 73, 80, 158, 169
α-linolenic acid, and ester 8, 9, 28, 158, 170
γ-linolenic acid, and ester 8, 9, 28–30, 65, 161, 208
linseed oil 62, 63, 68, 69
Linum usitatissimum 68
lipase 38
lipid absorption 224
lipid analysis 104
lipid classes 105, 106
lipid consumption 62, 226, 227
lipid digestion 224
lipid transport 224
lipids, definition of 1, 35
Lipofrac process 95
lipolysis 38, 120, 207, 224
lipoproteins 224, 225
lipoxygenase 182, 183
Lipozyme 208
long spacings, in long-chain compound unitcell 129
Lophira alata 6
Lophira procera 6
low resolution nmr spectroscopy 140
lubricants 239
Lunaria biennis 8, 79

macadamia oil 67, 78
maize oil 66

Mallotus phillipinensis 10
malonate 23
malvalic acid 13, 181
manaoic acid 14
margarine 97, 137, 232
marigold oil 80
mass spectrometry 147, 150, 152, 154
   MS–MS 154
MCT, *see* medium chain triglycerides
Mead's acid 8, 9, 28
meadowfoam oil 8, 80
medium chain triglycerides 56, 213, 224
melting behaviour 101
melting point 5, 93, 130, 132, 138
menhaden oil 72, 74, 174
mercury acetate 195
metal working fluids 239
metathesis 197
methanolysis 208, 209
methoxy mercuriacetates 196
methyl dihomo-γ-linolenate 30
methyl esters 208, 241
   use as solvent 241
methyl γ-linolenate 30
methylsterols 84
'methyl terrace' 135
methylmalonate 24
MGDG, *see* monogalactosyldiacylglycerols
milk fat 62, 71, 73, 229–232
minor components in oils and fats 82
molozonide 190
monkey nut oil 67
mono-acid triacylglycerols
   long spacings 135
   melting point 135
monoacylglycerols 37, 120, 210, 234, 235
   crystal structure 135
   derivatives 235
   synthesis 49, 55
monoene acids 6, 27
   melting point 132
   n-9 family 27
monogalactosyldiacylglycerols 36, 39
monoglycerides, *see* monoacylglycerols
Mucor javanicus 97, 213
Mucor miehei 97, 98, 208, 213
mustard seed oil 67, 79
mutton tallow 72
myristic acid 5
myrtle 178

near infrared spectroscopy 138
Nephelium lappaceum 6
nervonic acid 7, 8
neutral lipids 35, 107
neutralisation of a crude oil 89, 90
nickel catalyst 94
nicotinic ester 154

Niemann–Pick disease   44
nitrile   219, 220
nitrogen-containing compounds   201, 219
nmr spectroscopy
    ^{13}C   143, 146–148, 188
    ^{1}H   129, 140, 142, 143, 188
nomenclature, of fatty acids   1
non-ionic surfactants   205, 237
nylons   240

oats   178
octadecanoic acid   5
octadecatetraenoic acids   10
octadecatrienoic acids   10
octadecenoic acids   6, 7
    melting points   131
    synthesis   30
octadecynoic acids, melting point of   131
octanoic acid   5
octanol   14
octanone   14
odd acids, polymorphism   132
Oenothera biennis   65
Oenothera lamarkiana   65
oil
    content of plant sources   88, 100
    usage   63, 241
    yield   64
oil-bearing trees   61
oilfield chemicals   240
oilseed crops   77
Olea europea   68
oleic acid, and ester   6, 7, 28, 69, 73, 157,
    167
oleic oils   78
oleins   95
oleochemicals, annual production   230
oleodibehenin   55, 137
Olestra   57, 214
olive oil   62, 63, 68, 69, 74, 75, 79, 88, 120,
    174
    triacylglycerol composition   117
Omnium oxidative stability instrument
    104
oncobic acid   14
orange roughy oil   84
Oryza sativa   69
ox tallow   120, 124
oxidation   103, 162
    biological   180, 181
oxidative deterioration   103
oxidative fission   189
oxidative stability   103
oxidised fats   166
oxophytodienic acid   182
oxosphinganine   46
oxygenated acids   13
oxymercuration   195

ozonides   23, 191
ozonisation   191
ozonolysis   190, 192

PA, see phosphatidic acids
packet (stick) margarine   233
PAF, see platelet-activating factor
palm fruit   88
palm kernel oil   63
palm mid-fraction   96
palm oil   62, 63, 64, 68, 69, 75, 78, 120,
    126, 174
    tocols   85
palm olein   69, 95
palm stearin   69, 95
palm sterols   84
palmitic acid   5, 73
palmitoleic acid   6, 7
palm kernel oil   62, 63, 66, 68, 88, 174
    glycerol esters   114
paper making   239
parinaric acid   10
partial hydrogenation   22, 159
passionflower oil   67
PC, see phosphatidylcholines
PE, see phosphatidylethanolamines
peanut oil   67, 174
pecan oil   67
pentaerythritol esters   240
peracids, see peroxy acids, and esters
peroxide value   103, 104
peroxy acids, and esters   186, 215, 216
peroxyacetic acid   186
peroxyformic acid   186
petroselinic acid   7, 27, 69, 78, 79, 146
PG, see phosphatidylglycerols and also
    propyl gallate
phenolic antioxidants   178
phosphatidic acids   36, 39, 41
    synthesis   59
phosphatidylcholine, preparation   59
phosphatidylcholines   36, 40, 41, 81–83,
    122
phosphatidylethanolamines   36, 40, 41,
    81–83
phosphatidylglycerols   36, 40
phosphatidylinositols   36, 40, 81, 83
phosphatidylserines   36, 40
phosphoglycerides   36, 39
phospholipases   40, 41
phospholipids
    analysis   117
    sources, and composition   81
    synthesis   57, 58
phosphonic acid   40
phosphoric acid   179
photo-oxygenation   162, 164, 167
physical refining   89, 90

phytanic acid   12
phytol-based acids   12
phytosphingosine   44
PI, *see* phosphatidylinositides
picolinyl alcohol   151
picolinyl esters   151, 152
picolinyl oleate   152
pig fat   120, 124
pinolenic acid   10
pistachio oil   67
plasticisers   188
platelet-activating factor   42
polar lipids   35, 107
polyamides   200
polyene acids   8, 10, 27
   melting point   132
polyenes
   n-3 family   29
   n-6 family   29
polyethoxylated fatty acids   238
polyethyleneoxide amines   238
polyglycerol esters   236
polymethyl branched acids   12
polymorphism   129, 137, 138, 287
polyoxyethylene esters   222
polyurethanes   188
poppy seed oil   120
primrose oil, *see* evening primrose oil
pristanic acid   12
production (annual), oils and fats   61
production, of major oils   63
propanediol esters   43
propionate   24
propyl gallate   177
propylene glycol   236
prostacyclin   184
prostaglandins   29, 184
protecting groups   48, 49
PS, *see* phosphatidylserines
Pseudomonas sp.   210
psychosine   46
pumpkin oil   67
Punica granatum   10
punicic acid   10
pyrulic acid   11

quaternary ammonium compounds   219,
   220, 237

Raman spectra   138
rambutan tallow   6
Rancimat   103, 104
random distribution   211
randomisation   96
rapeseed oil   62, 63, 68, 69, 76, 78, 82, 84,
   88, 120, 124, 127
   tocols   85
rat liver phosphatidylcholines   122

refining   88, 89
regiospecific enzymes   212
R_f values   107
Rhizomucor miehei   207
Rhizopus arrhizus   97, 213
Rhizopus delemar   97, 213
Rhizopus niveus   97, 213
Ribes niger   65
rice bran oil   62, 69
rice bran wax   84
ricinoleic acid   13, 14, 69
Ricinus communis   65
Ritter reaction   203
rosemary   178, 179
rosmaric acid   179

safflower oil   69, 70, 79, 88, 147, 174
   triacylglycerol composition   17
sage   178
sal fat   66
salad oils   234
Salatrim   56
salmon phosphatidylcholines   122
Salva nilotica   180, 181
sampling   100
sand eel   72
santalbic acid   11
Sapium sebiferum seed oil   39
saponification   207
saponification equivalent   102
saponification value   102
sardine oil   72, 74
saturated acids   4
Schleichera trifuga   6
sebacic acid   14
selacholeic acid   7
selectivity factors   92
sesame oil   63, 67, 69, 70, 88, 174, 178,
   179
sesamol   178, 179
Sesamum indicum   70
shea butter   66
sheep tallow   120
Shorea robusta   66
Shorea stenoptera   66
short spacings, in long-chained compound
   unit cell   129
short-chain compounds   168
silica   90
silver ion chromatography   18, 111, 118
   hplc   119
silver ion systems   116
Simmondsia chinensis   8
single low rapeseed oil   68
sitosterol   84
slip melting point   101
sn, *see* stereospecifically numbered
soft (tub) margarine   233

solid fat content 102, 141
solid fat index 102
solvent crystallisation 95
sorbitan esters 236
sorbitan tristearate 234
soybean oil 62, 63, 69, 70, 76, 82, 88, 124, 174
  tocols 85
  triacylglycerol composition 117
soybean sterols 84
speciality oils 66, 67
sperm whale oil 84
sphinganine 46
sphingenine 44, 45, 47
  biosynthesis 46
sphingolipidoses 44
sphingolipids 43, 44
sphingomyelin 44, 47
  biosynthesis 47
sphingosine 36, 44
spreading fats 233
squalene 82, 83
stabilisers 188
Staphylococcus aureus 97, 213
steam deodorisation 196
steam refining 90
stearic acid 5
stearidonic acid 8, 29
stearins 95
stearolic acid 11
sterculic acid 13, 181
stereomutation 90, 92, 196
stereospecific analysis 122–127
stereospecific processes 118
stereospecifically numbered 35
sterol esters 36, 85
sterols 82, 84
stick (packet) margarine 233
stigmasterol 84
stillingia oil 39
Stokes aster 15, 81
Strophanthus seed oils 13
structured lipids 54
suberins 43
sucrose polyester 214, 225
sugar cane wax 84
sugar esters 36
sulphate 217
sulphonation 220
sulphur-containing compounds 202
sunflower oil 62, 63, 69, 70, 79, 82, 88, 174
  sterols 84
  tocols 84
Sunola 71
super refining 90
surface-active agents 237
surfactants 209, 234, 237

synlubes 240
synthesis
  fatty acids 45
  lipids 45
systematic names 1

tall oil 62, 69, 71, 78, 200
tallow 62, 63, 71, 78, 126, 174
  triacylglycerol composition 117
tallow-grease 63
tandem mass spectrometry 149, 153
tariric acid 10, 11
TBA test 103
TBHQ, *see* tertbutyl hydroquinone
TCL (triple chain length) structures 133, 136
tea 178
tertbutyl hydroquinone 177
tetracosanoic acid 5
tetracosenoic acid 7
tetradecanoic acid 5
tetrahydroxystearic acid 189
tetramethylalkanoic acids 12
textiles 239
Theobroma cacao 65, 66
thin-layer chromatography (tlc) 18, 19, 108
thiobarbituric acid test 104
three-quarter fat butter 232
three-quarter fat margarine 232
threshold flavour values 166
thromboxane 184
thyme 178
titre 101
tocols 83–85, 174, 178
tocopherols 82–85, 174, 178
tocotrienols 82–85, 174, 178
trans-acids 11, 102, 225, 229, 232
triacylglycerols 36, 37
  composition 74, 75, 76, 111, 114, 117
  synthesis 52, 55
trifluoroacetates 106
triglycerides, *see* triacylglycerols
trihydroxymethylpropane esters 240
trilaurin 134
trimethylalkanoic acids 12
trimethylsilyl ethers 106
triolein, hydrogenation 93
triterpenoid alcohols 85
tub margarine 233
tuna 74
tung oil 10, 62

ultraviolet spectroscopy 137
undecenoic acid 14
unit cell 130
unsaponifiable fraction 103
unsaturation 102

urea crystallisation   17
urethanes   123
urofuranic acids   16

vaccenic acid   7, 230
vegetable oils   69
vernolic acid   15, 81, 188
Vernonia galamensis   81
Vernonia sp.   15
visible fat   223
vitamin A   82
vitamin C   179
vitamin D   82
vitamin E   82, 84, 85, 175
vitamin K   82
VLDL (very low density lipoprotein)   225
von Rudloff oxidation   190

walnut oil   67, 174
    tocols   85
wax esters   36, 116
waxes   84
wheatgerm oil   67, 174
    tocols   85
Wightia seed oils   13
Wijs' reagent   102
winterisation   96
Wittig reaction   31
wool wax   84

X-ray diffraction   129
ximenynic acid   10, 11

Zea mays   66
zoomaric acid   6, 7